In Vitro Fertilization

The A.R.T. of Making Babies

Geoffrey Sher, M.D.
Virginia Marriage Davis, R.N., M.N.
Jean Stoess, M.A.

Facts On File®

AN INFOBASE HOLDINGS COMPANY

UNION COUNTY COLLEGE

3 9354 00118636 6

This book is dedicated to the millions of infertile couples in the United States, most of whom are economically disenfranchised from in vitro fertilization. We wrote In Vitro Fertilization *to help these couples (and the general public, too) understand that the medical technology necessary to enable them to have a baby is available today.*

In Vitro Fertilization: The A.R.T. of Making Babies

Copyright © 1995 by Geoffrey Sher, M.D., Virginia Marriage Davis, R.N., M.N., and Jean Stoess, M.A.

All rights reserved. No part of this book may be reproduced or utilized in any form or by any means, electronic or mechanical, including photocopying, recording, or by any information storage or retrieval systems, without permission in writing from the publisher. For information contact:

Facts On File, Inc.
460 Park Avenue South
New York NY 10016

Library of Congress Cataloging-in-Publication Data

Sher, Geoffrey, 1943–
    In vitro fertilization : the A.R.T. of making babies / Geoffrey
Sher, Virginia Marriage Davis, Jean Stoess.
        p. cm.
    Includes index.
    ISBN 0-8160-3269-6
    1. Fertilization in vitro, Human—Popular works.   2. Human
reproductive technology—Popular works.   I. Davis, Virginia
Marriage.   II. Stoess, Jean.   III. Title.
RG135.S543   1995
618.1′78059—dc20   95-2587

Facts On File books are available at special discounts when purchased in bulk quantities for businesses, associations, institutions or sales promotions. Please call our Special Sales Department in New York at 212/683-2244 or 800/322-8755.

Jacket design by Catherine Hyman

Illustrations by Marc Greene

MP FOF 10 9 8 7 6 5 4 3 2 1

This book is printed on acid-free paper.

Printed in the United States of America

CONTENTS

FOREWORD

Over the past 10 years, I have seen infertility from a variety of perspectives. As patient, as advocate, and as consumer protector, I have learned that the most critical element in a couple's journey through infertility is information. Yet for the millions of couples experiencing infertility, accurate, digestible data is difficult to find. Primarily because of this lack of information, I founded the de Miranda Institute for Infertility and Modern Parenting (dMI). Our organization strives to offer consumers the information they need to make crucial infertility decisions. In addition, I serve on a national committee for insurance advocacy that advises those who are interested in passing legislation mandating insurance coverage of in vitro fertility and gamete intrafallopian transfer.

For the past decade we have been answering questions about infertility, particularly questions about new reproductive technologies and insurance coverage. We rely on information we gather ourselves from many different sources. Multitudes of books exist on infertility, but none were specifically addressed to IVF and GIFT until the first edition of this book was written. All too frequently, even today, couples rely on information they glean from a few poor sources. Although dMI attempts to remedy this situation, what was really needed was a new book—and not just any book, but one that gradually builds consumer understanding of the process. That book is here, thanks to Dr. Geoffrey Sher and his coauthors, Virginia Marriage Davis and Jean Stoess. The authors have put together a definitive guide to the most intimate science: the new reproductive technologies. I am very happy to see this book updated, and I have complete confidence in its accuracy.

Dr. Geoffrey Sher is the executive medical director of the Pacific Fertility Medical Centers, a group of IVF programs located in California that have had enviable success. Most of the couples I know who have gone through IVF at these programs have felt that it was a positive experience regardless of outcome. I would have to concur as a former patient.

Infertile consumers think well of Pacific Fertility Medical Centers because the staff is caring and concerned. It is that feeling that motivated Dr. Sher to produce this book. Every page gives clear and concise information to infertile consumers starved for straight talk. Trying to understand what will, might, and could occur is sometimes frightening. But not knowing anything is worse. All too often, infertile couples go into treatment uncertain and unprepared. They are bewildered and frightened—hardly the state of mind associated with the joy of conception.

In this book, the most difficult aspects of treatment are stripped of jargon and served unadorned to the consumer. The authors explain how laboratories handle embryos. They also discuss the possibility of multiple births factually and clearly. Contrary to the belief among some infertile people that multiple pregnancies are an added bonus, they can be life-threatening. A patient's age, health, physique, and blood pressure are just some of the factors that must be considered in the event of multiple pregnancy. Decisions about multiple births must be made in advance. This consumer guide allows that kind of preplanning.

Finally, the section I counsel every infertile couple to read is the chapter on insurance coverage (Chapter 15). Pay attention to what the authors say and get involved. Political activism is the only way to get coverage. If you are not sure you deserve coverage, ask your insurance coordinator what happens to the maternity benefits you never use.

If you are infertile or considering GIFT or IVF, read this book. You need it. I hope that in a few years there will be no need for a consumer protection agency for the infertile. Until then, educate yourself.

Gina de Miranda
de Miranda Institute for Infertility and Modern Parenting
Orange County, CA

PREFACE

Innovators are rarely received with joy, and established authorities launch into condemnation of newer truths; for at every crossroad to the future are a thousand self-appointed guardians of the past.

—*Betty MacQuilty, "Victory Over Pain,"* Morton's Discovery of Anesthesia

In vitro fertilization (IVF) has come a long way since 1978, when Louise Brown, christened "the world's first test-tube baby" by the press, was born in England. The first in vitro fertilization program in the United States was introduced at the Eastern Virginia Medical School at Norfolk in the late 1970s. Now, more than 250 clinics throughout this country offer IVF, with varying degrees of reported success.

In vitro fertilization literally means "fertilization in glass." Traditionally known as in vitro fertilization and embryo transfer (IVF/ET), the procedure is more commonly referred to simply as in vitro fertilization, or IVF. (The term IVF will be used throughout this book instead of the more cumbersome IVF/ET.)

IVF comprises several basic steps. First, the woman is given fertility drugs that stimulate her ovaries to produce as many mature eggs as possible. Then, when the ovaries have been properly stimulated, the eggs are retrieved by suction through a needle inserted into her ovaries. The harvested eggs are then fertilized in a glass petri dish in the laboratory with her partner's or a donor's sperm. Several days later, the fertilized egg or eggs—now known as embryos—are transferred by a thin catheter through the woman's vagina into her uterus, where it is hoped they will grow into one or more healthy babies. Although IVF is an extremely promising procedure, it is no substitute for standard, less invasive (nonsurgical) methods for treating infertility. For this reason, it is essential that the infertile couple and their physician identify the cause of the infertility in order to determine the most appropriate form of treatment. This does not mean that IVF should be regarded as a treatment of last resort. It may well be that IVF offers the best hope for a healthy pregnancy. At most reputable IVF centers, the chance of a woman becoming pregnant with IVF can be much greater than that of a fertile woman conceiving (without treatment) in any given month of trying. Nevertheless, the couple should understand that IVF is not everything to everyone, and that some women will never get pregnant through IVF, no matter how many times they try.

Many infertile couples who have experienced repeated disappointments over the years in their attempts to conceive have come to our program in desperation. Most had previously tried a variety of unsuccessful procedures: fertility drugs for the woman and/or man, medications to treat various hormonal problems, nonsurgical alternatives such as artificial insemination, and pelvic surgery to repair anatomical defects. A few couples had been advised by their doctor to "take a holiday, just relax and get rid of your stress—and you'll get pregnant right away. Your problem's just emotional." All these couples looked to IVF as a promising new procedure that might help them conceive after all of their other attempts had failed.

Yet of the more than one million couples in this country for whom in vitro fertilization offers the best option for pregnancy, less than 20,000 couples undergo the procedure annually. Clearly, eligible infertile couples in the United States are not even coming close to tapping the potential of IVF. Why is this so?

One reason is that some people still consider IVF to be experimental. However, the evidence proves otherwise—more than 35,000 IVF babies have already been born in the United States. Yet the public, in company with many members of the medical profession, still knows surprisingly little about IVF beyond the often-sensational "test-tube baby" media coverage. One new father of an IVF baby explained how misleading that coverage can be:

People get a distorted idea about IVF when they turn on the TV and see a slide of a test tube with a baby inside. By no means are either a test tube or a baby involved at that point. There are just a momentary couple of days when fertilization takes place outside the couple's bodies, and then the embryo is placed in the woman's uterus and begins to grow there. The phrase "test-tube baby" is a convenient handle for the media, but it misleadingly implies a sterility that IVF doesn't have at all.

Unfortunately, consumers find it difficult to get much in-depth information about this exciting and relatively new procedure. (The term "consumers" is used here to mean both infertile couples and physicians who wish to refer their patients to a particular program.) Currently, no credible source provides prospectively audited, verifiable information about success rates obtained from IVF programs in the United States. As a result, people trying to learn about IVF often feel as though they are stumbling around in the dark.

Several national organizations, including the Society for Assisted Reproductive Technology (SART), an affiliated society of the American Society for Reproductive Medicine (ASRM), and a number of support groups for infertile couples, provide limited information about IVF and related procedures. SART was formed in 1988 under the umbrella of the ASRM, which is primarily made up of physicians but also includes laboratory personnel, psychologists, nurses, and other paramedical personnel interested in infertility.

The American Society for Reproductive Medicine/SART has taken the first step toward providing consumer-oriented information by compiling a registry of programs that have voluntarily submitted their IVF results from past years. SART will provide a list of IVF programs in the United States, but it does not recommend or endorse any programs. Instead, SART encourages consumers to contact the programs individually for more information. (The Society for Assisted Reproductive Technology, 1209 Montgomery Highway, Birmingham, AL 35216; telephone: [205] 978-5000, fax: [205] 978-5005.) (The ASRM was formerly known as the American Fertility Society.)

Until 1990, ASRM lumped all data from reporting programs together and derived an overall IVF success-rate statistic for the United States. Since then, under pressure from Congress, SART has required clinic-specific data from all its members. Not all clinics are members of SART, however. An additional problem is that the data SART currently provides are voluntarily submitted in an unaudited fashion by member programs and are accordingly unverifiable. Being retrospective, these data have many flaws.

In 1993, SART contracted with a nationally respected auditing firm to initiate a process of data collection that would ultimately lead to a clinic-specific third-party review of IVF outcome statistics. The process of data collection was introduced and refined in 1994 in the hope of implementing the reviewing process of all SART-affiliated IVF programs for those procedures performed during 1995. When this book was being written, SART's timeline for producing a monitored report allowed for at least a nine-month period between the time of data collection and the time of collation and reporting. This means that a third-party-reviewed IVF outcome report is unlikely to be available to the consumer until at least 1997, a year after the statistics have been gathered.

As of September 1992, Pacific Fertility Medical Center has required all of their programs to submit information about every patient undergoing

IVF, prior to the egg-retrieval procedure, to a nationally respected auditing firm, where the information is compiled and released intermittently as a verifiable independent audit. (For privacy purposes, the couples whose statistics are reported to the auditor are not identified.) Pacific Fertility Medical Centers is currently the only group of programs in the United States that has voluntarily submitted to this degree of accountability. We will discuss the importance of this action, as well as accountability, in detail in Chapter 15.

A number of fertility support groups in the United States also provide information about IVF and about infertility in general. The largest group is Resolve, Inc., a national nonprofit organization that offers counseling, referral services, and support to infertile couples (1310 Broadway, Somerville, MA 02144; telephone: [business office] [617] 623-1156; [help line]: [617] 623-0744). Resolve has local chapters throughout the country, and publishes a bimonthly newsletter and other literature about infertility. Resolve and several of the smaller support groups also provide lists of IVF programs, but they are reluctant to recommend specific programs.

Aside from the problem of insufficient information, another obstacle to widespread acceptance of IVF is its high cost. In vitro fertilization is relatively expensive ($7,000–$10,000 per procedure depending on the program). Some insurance companies currently fund about one-third to one-half of the total cost of IVF (the fertility hormone shots, ultrasound examinations, and the ultrasound egg-retrieval procedure, including operating room and anesthesiologist's fees). But with few exceptions, they still will not pay for laboratory work, the fertilization process, or the embryo transfer.

Many couples pass up IVF because of the financial burden, although it may be the most appropriate treatment for them. They simply can't afford it. Some states have passed laws requiring insurance companies to reimburse in total for IVF, and several others are considering similar legislation. Nevertheless, IVF will continue to be beyond the financial reach of most couples in this country until insurance companies adopt equitable reimbursement policies.

We must warn consumers that the outlook on IVF-related issues is not likely to improve for some time. Information will continue to be difficult to obtain, and IVF will remain an expensive procedure. But by researching the IVF situation for themselves, couples will be able to answer these fundamental questions: (1) Are we eligible for IVF? and (2) How do we select the program that will give us the best results?

This book has been designed to help answer these critical questions. It describes IVF and some other Assisted Reproductive Technology (ART) procedures; outlines a variety of emotional, physical, financial, and moral/religious issues; and highlights points that should be considered when deciding whether IVF or another high-tech procedure is indeed the most appropriate option. We do not offer any judgments relating to ethics, religion, or morality. These kinds of decisions are private matters that must be resolved by each couple in their own way. We do not intend to imply that our approach is the only acceptable way and/or should be rigidly adhered to. Our function is to recommend, to inform, to educate, and to serve—but never to dictate.

We are particularly cognizant of the fact that many women who do conceive following IVF may have pregnancies that are at risk. Going from infertility to family can be extremely traumatic from an emotional, psychological, and physical point of view. One of the ways we prepare couples in our setting is by providing them with as much information about infertility, as well as IVF and related procedures (ART), as they need. We believe that being as knowledgeable as possible helps them cope with the roller-coaster experience they will undergo.

Accordingly, we will be glad to provide the readers of this book any information they might request about infertility. We would also be happy to send information regarding the appropriate specialists in high-risk obstetrics in the readers' geographic area. To request this material, call our Patient Resource Center at 1-800-999-9075.

Finally, we wrote this book to help consumers develop and maintain realistic expectations about IVF. Realistic expectations revolve around the best and the worst possible scenarios, but all infertile couples should prepare themselves for the worst, just in case. However, by planning an effective strategy, asking the right questions, and evaluating the answers properly, candidates can determine whether they are eligible for IVF and can find the most appropriate program. As a 35-year-old new mother told us, doing that homework does pay off:

I must have spent at least three hours talking with my own physician trying to find out where to go for IVF. It was so frustrating not being able to find anyone who could give me any real answers. Several times I was tempted to go to the IVF clinic nearest us just because it was so convenient. But, thank God, I did my homework as thoroughly as I knew how. I must have called up fifteen different programs. I asked a lot of questions about success rates and what it was like to go through their programs, and then I had to sort everything out. I finally found a great program—and now we have a beautiful little girl who is the joy of our lives. I can hardly remember what life had been like without her. It was worth all that effort.

Geoffrey Sher, M.D.

1

THE GROWING DILEMMA OF INFERTILITY

It is estimated that there are about 40 million couples of childbearing age living together in the United States today. Approximately 3.3 million of these couples are infertile.

This estimate is based on a series of nationwide surveys conducted by the National Center for Health Statistics. In a 1982 report that surveyed cohabiting married couples in which the women were between 15 and 44 years of age, it was found that one out of every 12 couples, or 8.5%, were involuntarily infertile. Another survey, released relatively recently, revealed similar results. The figure of 3.3 million infertile couples was derived by adjusting the percentage slightly upward to allow for unmarried cohabiting couples.

Infertility can be defined as the inability to conceive after one full year of normal, regular heterosexual intercourse without the use of any contraception. The odds that a woman will get pregnant without medical assistance when she has failed to do so after a year or two of unprotected intercourse are extremely low.

Only couples who have experienced the problem of infertility can truly understand its devastating emotional and physical impact. As one woman who had been trying unsuccessfully to become pregnant for many years explained:

It has been two years since I learned the reason I wasn't getting pregnant was because my fallopian tubes were blocked. It is incredible to think that I have had more physical assault on my body in the last two years than in the rest of my thirty years combined. I underwent it voluntarily, too, because I wanted to correct the problem and have a child. Yet, all the surgeries, the tests, and the medications seemed relatively minor compared to the emotional burden I put on myself.

After my first surgery, when my physician said it was okay to try to become pregnant, I don't think there was ever a day, or perhaps even an hour, that I didn't think about conceiving. It was always there—when I would see a child in the grocery store. When my friends would gripe about their kids. When I was on day 1, or day 14, and every other day of my menstrual cycle. Whenever my husband and I made love. Was I ever going to get pregnant? I had conflicting fantasies of what I would be like as a 60-year-old woman who had never had children. I could never get it out of my mind.

One study of infertile couples illustrates the pervasive impact of infertility. When asked what they considered to be the primary problem in their lives, almost 80% of the couples replied that it was their inability to conceive. Most of the remaining 20% ranked infertility as their second most perplexing problem, after financial difficulties. The remaining fraction of respondents rated infertility a close third after financial problems and marital strife.

One newly pregnant woman, who had just completed her second IVF treatment cycle, summed up the emotional impact of infertility in this way:

You really can't understand what it's like to be infertile unless you are infertile yourself and have experienced what we've gone through. You can sympathize, but you can't empathize with us.

The traditional options available to infertile couples who want a baby have included counseling, surgery to repair anatomical damage, the use of fertility drugs to enhance ovulation and sperm function, and insemination of the woman with her partner's or a donor's sperm. Most authorities would agree that these methods are effective for approximately 50% of all infertile couples.

For the remaining 50%, or about 1,600,000 couples annually in the United States, the only recourse is IVF—yet fewer than 40,000 IVF procedures are being performed annually. Given that the majority of these couples would require two or more attempts at IVF to have a baby, this means that fewer than 20,000 couples actually undergo the procedure annually.

WHY THE NUMBER OF INFERTILE COUPLES IN THE UNITED STATES IS INCREASING EVERY YEAR

According to the National Center for Health Statistics, the rate of involuntary infertility has remained constant at about 8.5% since 1965. This is an increase over the figures for the early 1950s through 1964, when it was thought that approximately 7 to 8% of all couples were unable to conceive. Although the rate of infertility has not changed since the mid-1960s, the number of infertile couples has increased every year in step with population growth. The following factors contribute to this trend.

Venereal Diseases Are Epidemic in the United States

Once considered largely under control because of the discovery and availability of proper medication, venereal diseases that damage the reproductive system are on the rise again. One of the major causes of this is the increased availability of effective birth control methods, which has undoubtedly contributed to a more open approach toward sexual activity. Consequently, both men and women often have relatively large numbers of sexual partners. The unfortunate result has been a significant increase in sexually transmitted diseases.

Venereal diseases such as *gonorrhea* are rampant in the United States. Gonorrhea lodges in the woman's fallopian tubes and often results in a severe illness. Unfortunately, in many cases gonorrhea causes so little physical discomfort that women frequently do not bother to seek treatment. But even a minor gonorrheal infection can damage the fallopian tubes, and many women find out they have had gonorrhea only when they investigate the cause of their infertility. In contrast, gonorrhea in men almost always produces sudden, painful symptoms that usually prompt an immediate visit to the doctor. For this reason, men are less likely to be infertile due to gonorrhea than are women.

The cure of sexually transmitted diseases is complicated by the emergence of new strains of gonorrhea and other venereal diseases that resist traditional treatment. These bacteria can be combatted only by new and expensive antibiotics that often are not readily available.

Chlamydia, another infection that damages and blocks the fallopian tubes, is more common today than gonorrhea. Chlamydia is relatively difficult to diagnose and culture, and it responds well only to specific antibiotics. Past infection is diagnosed through a specific staining procedure of cervical secretions or by blood testing for antichlamydia antibodies. The symptoms of chlamydia are very similar to those of gonorrhea.

Syphilis, too, is again becoming widespread in the United States. Syphilis is easily cured in its early stages. In later stages its spread can be halted, but its effects cannot be reversed.

The Biological Clock Keeps On Ticking

Many women today choose to delay beginning a family until they are at least 30—in order to establish their careers, to be sure they and their partner can afford it, because they married relatively late, or because they want to have a child with a new partner. However, there is sometimes a price to pay for having children later on. For example, some disorders that produce infertility tend to appear during the second half of a woman's reproductive life span. Thus, a woman who decides to have children after 35 might find she is infertile because of hormonal problems, a pelvic disease such as endometriosis, or the development of benign fibroid tumors of the uterus. In addition, the ability to ovulate healthy eggs and concurrently generate a hormonal environment that can adequately support a pregnancy becomes increasingly compromised as a woman gets older. Thus, many women who plan to become pregnant later in their reproductive lives find themselves unable to do so.

Medications and "Recreational Drugs" Are Taking Their Toll

Alcohol, nicotine, marijuana, cocaine, and other psychotropic drugs can significantly reduce both male and female fertility because they are capable of altering the genetic material of eggs and sperm. However, these substances can potentially have a far more severe and long-lasting effect on the woman than on the man. Because a woman is born with her lifetime quota of eggs already inside her ovaries, unwise use of medications and drugs can damage all the eggs her body will ever produce. In contrast, a man generates a completely new supply of sperm approximatelyevery three months, so damaged sperm are replaced in a short time (see Chapter 2).

Although the infertility rate has not increased, the aggregate number of couples whose last option for pregnancy is one of the sophisticated new ART procedures is skyrocketing. Recent advances in the evolution of high-tech methods to evaluate and treat infertility offer the hope of pregnancy to couples who had no hope until now. In vitro fertilization is just one of these promising procedures, but in many cases it offers the best hope for success. To date more than 35,000 very special miracles have occurred in the United States: the gift of life through in vitro fertilization has been given to thousands of infertile couples who would otherwise have been childless.

CHAPTER

2

THE ANATOMY AND PHYSIOLOGY OF REPRODUCTION

In vitro fertilization can be viewed as an extension of the normal human reproductive process. It merely bypasses many of the anatomical or physiological causes of infertility by substituting IVF techniques for some of the processes that occur naturally in the body. In order to understand both natural conception and IVF, therefore, one must first be familiar with human reproductive anatomy and the process of reproduction.

THE FEMALE REPRODUCTIVE TRACT

The female reproductive tract consists of the vulva, vagina, cervix, uterus, fallopian tubes, and ovaries. The external portion of the female reproductive tract (see Figure 2-1) is known as the *vulva*. The vulva includes the inner and outer lips, or *labia*. The hair-covered outer labia are called the *labia majora* (major lips). The *labia minora*, smaller inner lips partially hidden by the labia majora, are remnants of tissue whose embryologic counterpart in the male develops into the scrotum.

The *clitoris*, a small organ at the junction of the labia minora in the front of the vulva, is the embryologic counterpart of the male penis. The

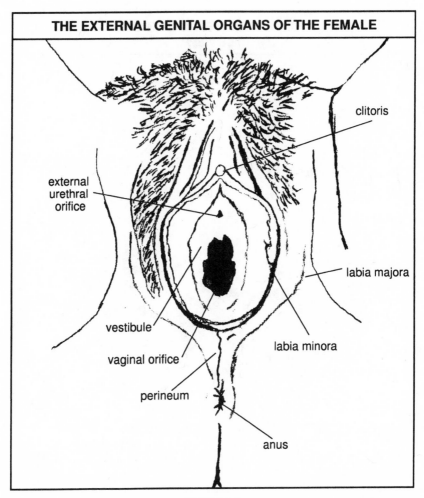

THE EXTERNAL GENITAL ORGANS OF THE FEMALE

clitoris

external
urethral
orifice

labia majora

vestibule

labia minora

vaginal orifice

perineum

anus

FIGURE 2-1

clitoris undergoes erection during erotic stimulation and plays an important role in orgasm.

The area between the labia minora and the anus is called the *perineum*. It is formed by the outer portion of the fibromuscular wall and skin that separate the *anus* and *rectum* from the vagina and vulva.

The *vagina*, a narrow passage about 3 ½ to 4 inches long and about 1 inch wide, spans the area between the vulva and the cervix. It opens outward through the cleft between the labia minora, or *vestibule*. The vagina's elastic tissue, muscle, and skin have enormous ability to stretch so as to accommodate the penis during the sex act and the passage of a

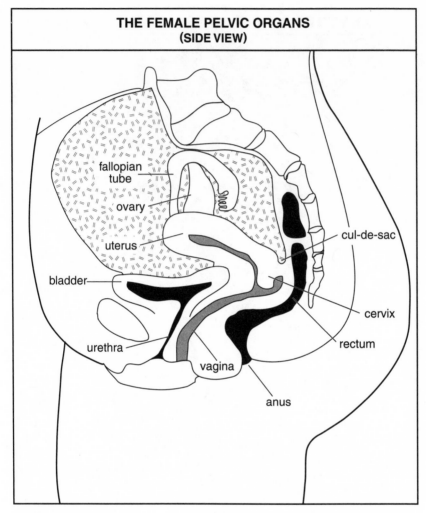

FIGURE 2-2

baby at birth. The vagina is actually a potential space; it is a real space only when the penis enters it or during childbirth. At other times, the vaginal walls are collapsed against one another; a cross-section of a relaxed vagina would resemble the letter H. In front of the vagina lie the *bladder* and the *urethra* (outlet from the bladder), and at the back is the rectum (see Figure 2-2).

The *cervix,* which is the lowermost part of the uterus, protrudes like a bottleneck into the upper vagina. As Figure 2-3 illustrates, a *fornix,* or deep recess, is created around the area where the cervix extends into the

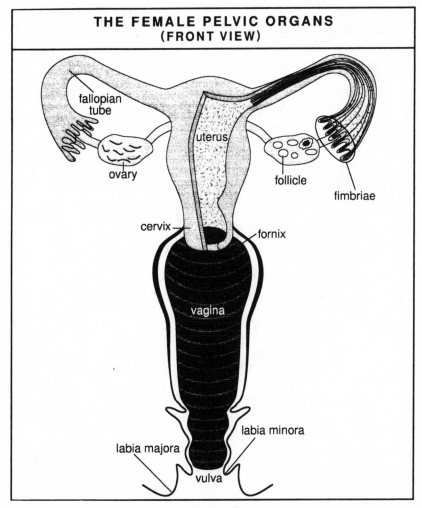

**THE FEMALE PELVIC ORGANS
(FRONT VIEW)**

fallopian tube

uterus

ovary

follicle

fimbriae

cervix

fornix

vagina

labia minora

labia majora

vulva

FIGURE 2-3

vagina. The area of the abdominal cavity behind the uterus is known as the *cul-de-sac*. The cervix opens into the uterus through a narrow canal, the lining of which contains glands that produce cervical mucus (the important role that cervical mucus plays in the reproductive process will be explained later). The cervix is particularly vulnerable to infections and other diseases such as cancer.

The *uterus*, which consists of strong muscle fibers, is able to stretch and grow from its normal size, when it resembles a pear, to accommodate a full-term pregnancy. The valvelike transition between the cervix and the

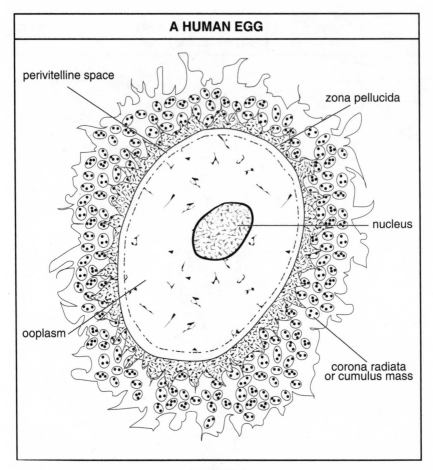

A HUMAN EGG

perivitelline space

zona pellucida

nucleus

ooplasm

corona radiata
or cumulus mass

FIGURE 2-4

uterine cavity enables a baby to grow within the uterus without prematurely dilating the cervix and thereby endangering the pregnancy through miscarriage or premature birth. The lining of the uterus, which nurtures and supports the developing embryo, is known as the *endometrium*.

The *fallopian tubes* are two narrow 4-inch-long structures that lead from either side of the uterus to the ovaries. At the end of the fallopian tubes are fingerlike protrusions known as *fimbriae*.

The *ovaries* are two almondlike structures attached to each side of the pelvis adjacent to the fimbriae. The ovaries both release eggs and discharge certain hormones into the bloodstream. The process of releasing the egg or eggs is called *ovulation*. Eggs are also known as *ova* or *oocytes*.

About the size of a grain of sand, eggs are the largest cells in the human body. (A woman develops all the eggs she will ever have at the fetal age of 12 weeks.) Each month the ovaries select a number of the woman's lifetime quota of eggs for maturation. However, only one egg, and sometimes two, actually reach the stage where they are mature enough to be released and possibly fertilized. Eggs that do not mature are absorbed by the ovaries after ovulation.

Although a female baby starts off with about seven million eggs when she is inside her mother's womb, her ovaries contain only about 700,000 eggs by the time she reaches puberty. A woman uses about 300,000 of these eggs during the approximately 400 ovulations that occur during her reproductive life span.

Eggs mature in blisterlike structures, or *follicles*, that project from the surface of the ovaries. At ovulation, the egg is not simply expelled into the abdominal cavity. Instead, the fimbriae at the ends of the fallopian tubes gently vacuum the surface of the ovaries to retrieve the egg and direct it through the fallopian tubes for possible fertilization.

The human egg (see Figure 2-4) is similar in structure to the eggs of many other species, including the chicken. In the center of the human egg is the *nucleus*, which bears the chromosomes. The surrounding *ooplasm* contains *microorganelles*, which are cellular factories that produce energy for the egg. The ooplasm also contains nurturing material that supports the embryo during its early stages after fertilization, thus enabling it to grow before becoming attached to an external source of nourishment. Surrounding the ooplasm and nuclear material is the *perivitelline membrane*, which separates the internal matter from the *zona pellucida*. The zona pellucida is analogous to the shell of a chicken egg, and the perivitelline membrane corresponds to the membrane inside the eggshell. The human egg, unlike the chicken egg, also contains a group of cells known as the *cumulus granulosa*, which are arranged in a starburst effect around the outside of the zona pellucida. (The critical role each of these structures plays in the fertilization process is explained later in this chapter under "How Fertilization Occurs.")

THE MALE REPRODUCTIVE TRACT

The male sex organs (see Figure 2-5) comprise the *penis* and two *testicles*, or *testes*, which are located in a pouch called the *scrotum*. The testicles (male counterparts of the woman's ovaries) produce *spermatozoa*, or *sperm* (the male equivalent of the woman's eggs).

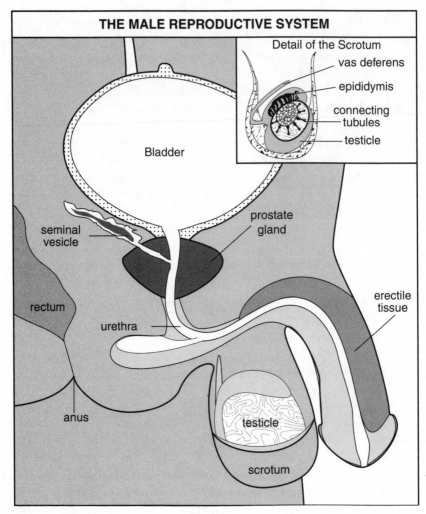

FIGURE 2-5

In contrast to the woman, who is born with a lifetime quota of eggs, the man's testicles generate a new complement of sperm approximately every 100 days. The sperm begin to mature in the testicles and continue to develop as they travel through a long, thin, coiled tubular system in the scrotum called the *epididymis.* The epididymis is connected to a straight, thicker tube called the *vas deferens.* Just before the vas deferens enters the penis it joins the *urethra,* which originates in the bladder and allows the passage of urine from the bladder through the penis. Sperm are transported through this system by muscular contractions known as *peristalsis.*

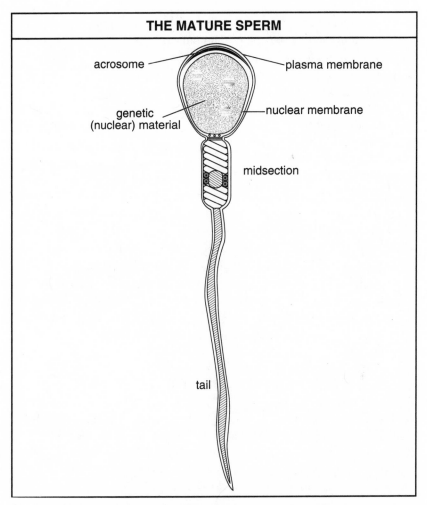

FIGURE 2-6

Several glands, including the *seminal vesicles* and the *prostate gland,* are located along this tract. They release a large amount of milky secretions that nurture and promote the survival of the sperm. The combination of sperm and milky fluid that is ejaculated during erotic experiences is known as *semen.* Semen and urine are not discharged simultaneously through the urethra. Urine is prevented from mixing with semen in the urethra because the bladder-urethra opening constricts during ejaculation. Similarly, closure of the vas deferens–urethra juncture prevents passage of semen during urination. In certain cases, removal of a diseased prostate gland may compromise this separation effect and cause the man to ejaculate backward into the bladder rather than outward through the penis; this is known as *retrograde ejaculation*. This condition may cause infertility, but it can be treated by inseminating the woman with sperm separated from urine the man would pass immediately following orgasm.

Sperm (see Figure 2-6), in contrast to eggs, are the smallest cells of the body. They resemble microscopic tadpoles. Each sperm consists of a head, whose *nucleus* contains the hereditary or genetic material arrayed on chromosomes, a midsection that provides energy, and a tail that propels the sperm along the male reproductive system and through the woman's reproductive tract. The top of the head is covered by the *acrosome*, a protective structure containing enzymes that enable the sperm to penetrate the egg; and the surface of the acrosome is enveloped by the *plasma membrane*. (The function of the acrosome and plasma membrane will be explained under "How Fertilization Occurs" later in this chapter.)

Eggs and sperm are called *gametes* until fertilization. A fertilized egg is called a *zygote* until it begins to divide; from initial cell division through the first eight weeks of gestation it is known as an *embryo*, and from the ninth week of gestation until delivery it is called a *fetus*.

HOW THE GENETIC BLUEPRINT IS DRAWN

Each cell in a human being (except for the *gametes*, or eggs and sperm) contains 46 chromosomes, which are bound together into 23 pairs. Chromosomes contain hundreds of thousands of genes, each of which transmits the hereditary messages of the man or woman.

If a sperm containing 46 chromosomes were to fertilize an egg that also contained 46 chromosomes, it is obvious that the two gametes would produce a zygote containing double the proper number of chromosomes. Therefore, nature has created a method through which the number of chromosomes in both the sperm and the egg are reduced by half. This reduction-division, which occurs immediately prior to and during fertilization, is referred to as *meiosis* (see Figure 2-7).

As a result of this process, a newly fertilized zygote also contains 46 chromosomes. All that has been exchanged is the chromosomal material, including the genes. This simple-looking cell is actually quite complex, for within its boundaries lies the information for 100,000 chemical reactions: the blueprint for a new individual. The set of instructions carried on the chromosomes is complete, and a virtual explosion of embryonic development is imminent.

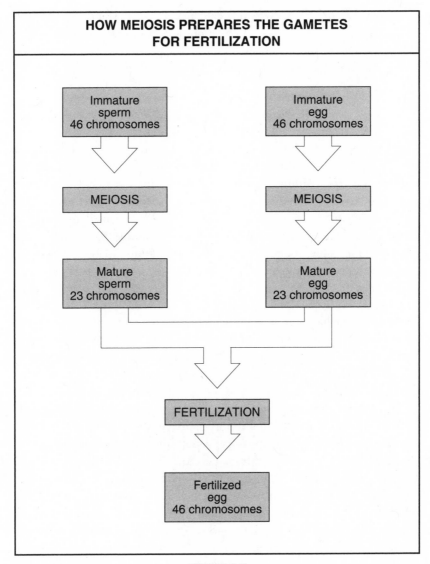

FIGURE 2-7

When the cells of the zygote begin to divide, and continue to divide over and over, they replicate the same number of chromosomes that occurred in the zygote. Thus, every new cell has 23 identical pairs of chromosomes (46 total) in its own new image. Such division for the purpose of replicating cells identically is called *mitosis*. The growth and development of all tissues—with the exception, of course, of the gametes—is done by mitosis.

Obviously, such a sensitive, intricate mechanism can and often does go wrong. Because this process malfunctions so often, nature has devised a way of selecting out its mistakes and discarding them. This will be discussed later in this chapter under "Miscarriage in Early Pregnancy."

THE PROCESS OF FERTILIZATION

Fertilization is a complex process that must be accomplished within a strict time frame. Theoretically, a man is always fertile, but a woman's egg can only be fertilized within a specific 12- to 24-hour period shortly after ovulation. Therefore, there is a "window of opportunity" lasting only 24 to 48 hours each month when intercourse can be expected to result in fertilization. Timing is critical if the egg and sperm are to survive the journey through the woman's reproductive tract, unite, become fertilized, and result in the embryo implanting successfully into the uterine wall.

The man deposits between 100 and 200 million sperm into the woman's vagina with each ejaculation of semen. During normal intercourse, or even after the woman has been artificially inseminated, much of the semen pools in the posterior fornix behind the protruding cervix. Because the cervix usually points partially backward into the posterior fornix, the cervix is usually immersed in the pool of ejaculated semen. This immersion helps direct the sperm through the cervix and into the reproductive tract.

The journey from the fornix to the fallopian tubes, which is about four inches, is hazardous and unbelievably taxing for the tiny sperm. For a cell the size of a sperm to travel this distance is equivalent to an adult human swimming the Pacific Ocean from Los Angeles to Tahiti. Only a small fraction of exceptionally strong, healthy sperm out of several million that were deposited in the fornix will survive that 24- to 48-hour journey. Many are killed by the hostile environment in the vagina or cervix, and others simply do not survive the long swim. Peristaltic contractions in the fallopian tubes help the remaining sperm reach the egg, and the same contractions propel the fertilized egg or embryo back through the fallopian tube to the uterus. Amazingly, out of the millions of sperm ejaculated into the vagina, only a few hundred to a few thousand successfully complete the journey to the waiting egg in the fallopian tube.

An egg presents a large target for the tiny sperm: about $\frac{1}{180}$ inch as opposed to the sperm's $\frac{1}{100,000}$ inch. This means that the egg is about 550 times as wide as the sperm. The difference in size between the two gametes is due to the massive amount of cytoplasm within the egg that will nourish the newly formed embryo. Sperm, in contrast, consist almost entirely of genetic material with very little cytoplasm. They are little more than bags of chromosomes propelled by a tail.

In vitro fertilization improves the odds that sperm can find and fertilize an egg. This is because the distance sperm have to swim to find the egg in a petri dish is considerably shorter than the "long-distance route to Tahiti" found in nature. The egg still lies passively within the dish, as it would in the fallopian tube, but in the significantly reduced volume of the dish the sperm are far more likely to find the egg than they would be if they had to negotiate the entire distance from the fornix.

How Fertilization Occurs

The process whereby sperm are prepared to fertilize an egg, a process known as *capacitation,* takes place in two stages. First, as a sperm passes through the woman's reproductive tract its acrosome fuses with its plasma membrane, slowly releasing the enzymes within the acrosome. Then, with its acrosome now exposed, the sperm attacks the cumulus-granulosa and the zona pellucida of the egg. The heads of a number of sperm fuse with the zona pellucida, and one successful sperm penetrates the egg. The process whereby a sperm fuses with the zona, the second stage of capacitation, is called the *acrosome reaction.* The sperm require from 5 to 10 hours of incubation in the fluids of the female reproductive tract to complete the acrosome reaction.

Capacitation takes place in the mucus secretions of the cervical canal, and continues in the uterus and fallopian tubes. It is believed that the passage of sperm through the cervical mucus around the time of ovulation promotes the necessary physical, chemical, and structural changes in the plasma membrane to facilitate release of acrosomal enzymes. (Because only sperm that have undergone capacitation are able to fertilize an egg, in IVF therapy the first stage of capacitation must be replicated in the laboratory prior to IVF if fertilization is to occur in the petri dish.)

After the acrosome reaction has taken place, the successful sperm completes the fertilization process by burrowing through the zona pellucida (the egg's shell-like covering) and ooplasm to the nuclear material. The sperm sheds its body and tail upon penetration, and only the head (containing the genetic material) actually enters the egg. Figure 2-8 illustrates the following steps in the capacitation-fertilization process:

Phase 1: The plasma membrane fuses with the acrosome as the sperm pass into the reproductive tract to reach the egg, thus initiating capacitation.

Phase 2a: The acrosomal enzymes are released and penetrate the cumulus mass (corona radiata) cells of the egg.

Phase 2b: The acrosome fuses with the zona pellucida, thus completing capacitation.

Phase 3:   The sperm burrows through the zona pellucida into the ooplasm of the egg.

The moment a sperm penetrates the egg's zona pellucida, a reaction in the egg fuses the zona and the perivitelline membrane into an impermeable shield that prevents other sperm from entering. The entry of more than one sperm into a fertilized egg (called *polyspermia*) causes the resulting embryo to die.

When fertilization occurs, the egg starts dividing within the zona covering, drawing its metabolic supplies from the ooplasm within the egg. Propelled by contractions of the fallopian tube, the dividing embryo begins its three- or four-day journey back to the uterus and continues to divide after it reaches the uterus. (The fertilization process occurs near the middle of the fallopian tube—not in the uterus.)

About two days after reaching the uterus, when the embryo has divided into about 30 cells, it cracks open, and all the cells burst out through the

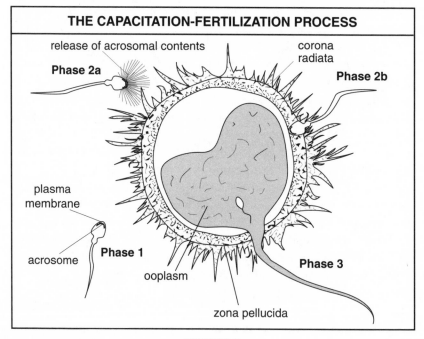

FIGURE 2-8

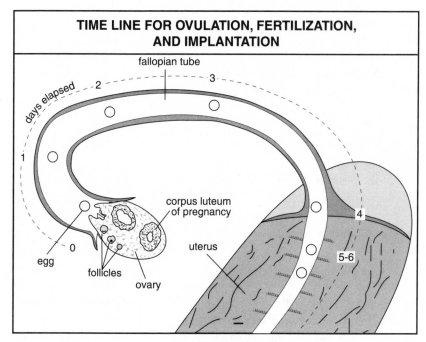

FIGURE 2-9

fractured zona. This is known as *hatching*. These cells then try to burrow their way into the lining of the uterine wall. A portion of the growing embryo soon makes contact with the mother's circulatory system and becomes the earliest form of the placenta, from which the baby will receive its nourishment (see Figure 2-9).

If an embryo implants anywhere but in the uterus it is referred to as an *ectopic pregnancy.* In most cases an ectopic pregnancy is due to embryonic implantation in the fallopian tube; this occurs about once in every 200 pregnancies. Isolated cases of implantation in the reproductive tract, such as on the ovary or elsewhere in the abdominal cavity, have been reported. In rare instances such ectopic pregnancies have been known to develop to full term, but the baby invariably will not survive.

Figure 2-9 follows the progress of an egg as it is ovulated from the follicle, becomes fertilized in the fallopian tube, and implants into the endometrium of the uterus. The dotted line plots the days that normally elapse as

1. ovulation occurs (and meiosis takes place prior to and during fertilization)

2. the fertilized egg, which has not yet divided, is now known as a *zygote*

3. the egg begins to divide and is now known as an embryo; at this point each *blastomere*, or cell, within the embryo is capable of developing into an identical embryo

4. the embryo develops into a mulberry-like structure known as a *morula*

5–6. a cavity develops within the embryo, which indicates the *blastocyst* stage

7. the process of *gastrulation* begins (cells are now dedicated to the development of specific embryonic layers that subsequently will form specific organs and structures; individual cells are no longer capable of developing into embryos).

At ovulation, the physical-chemical properties of the cervical mucus nurture the sperm as they pass through it, enhancing their quick passage and therefore capacitation as well. This is because hormonal changes around the time of ovulation ensure that the microfibrilles, or *myceles*, of the cervical mucus are arranged in a parallel manner. The sperm must then swim between the myceles in order to reach the uterus and finally the fallopian tubes. In addition, the cervical mucus becomes watery, and the amount produced (some of which may be discharged) increases significantly.

At other times during the menstrual cycle the hormonal environment alters the arrangement of the myceles in the cervical mucus to form a barrier to the passage of sperm. During this time the mucus is thick, thus preventing the sperm from passing through the cervix.

The *Billings Method* of contraception is based on this phenomenon. A woman using the Billings Method predicts when she is likely to be ovulating by evaluating whether her cervical mucus is thick or watery. Because pregnancy can occur only around the time of ovulation, it is accordingly possible for her to identify the so-called "safe period" when she is unlikely to conceive following unprotected intercourse.

HORMONES PREPARE THE BODY FOR CONCEPTION

Pregnancy, of course, begins with the fusion of two gametes—the female egg and the male sperm—but the preparations for conception begin long before fertilization occurs. The onset of puberty in both the man and the woman sets the stage for a biorhythmical hormonal orchestration that becomes more and more fine-tuned over the ensuing decade. It begins with the formation and release of hormones into the bloodstream, and the bodies of both sexes rely on a complex feedback mechanism to measure existing hormonal levels and determine when additional hormones should be released.

The *hypothalamus* (a small area in the midportion of the brain) and the *pituitary gland* (a small, grapelike structure that hangs from the base of the brain by a thin stalk) together regulate the formation and release of hormones. The hypothalamus, through its sensors, or "receptors," constantly monitors female and male hormonal concentrations in the bloodstream and responds by regulating the release of small proteinlike "messenger hormones" to the pituitary gland. These messenger hormones are known as *gonadotropin-releasing hormones,* or GnRH (see Figure 2-10).

In response to the messenger hormones from the hypothalamus, the pituitary gland determines the exact amount of hormones that it in turn will release to stimulate the *gonads* (ovaries in the woman and testicles in the man). These hormones, called *gonadotropins,* are FSH (*follicle-stimulating hormones*) and LH (*luteinizing hormones*). The hypothalamus closes the feedback circle by measuring the level of hormones produced by the gonads while at the same time monitoring the release of LH and FSH by the pituitary gland.

This "push-pull" interplay of messages and responses produces the cyclical hormonal environment in the woman that is designed solely to promote pregnancy. A "push-pull" mechanism also occurs in the man with regard to the release of testosterone. Similar feedback mechanisms regulate other hormonal responses in mammals, such as the functioning of the thyroid and adrenal glands.

There are two primary sex hormones in the female, *estrogen* and *progesterone;* in the male there is only one, *testosterone.* The pituitary gland releases identical hormones—FSH and LH—to the gonads of both the woman (the ovaries) and the man (the testicles), but the female and male gonads respond differently to these hormones. The level of female hormones fluctuates approximately monthly throughout the menstrual cycle, while male hormone production remains relatively constant.

In men, FSH and LH trigger the production of testosterone and influence the production and maturation of sperm. (The mechanism of male hormone production is not relevant to a proper understanding of IVF and will not be discussed in detail here.)

In women, extraneous factors as well as the level of circulating hormones and gonadotropins may influence the body's feedback mechanism. For

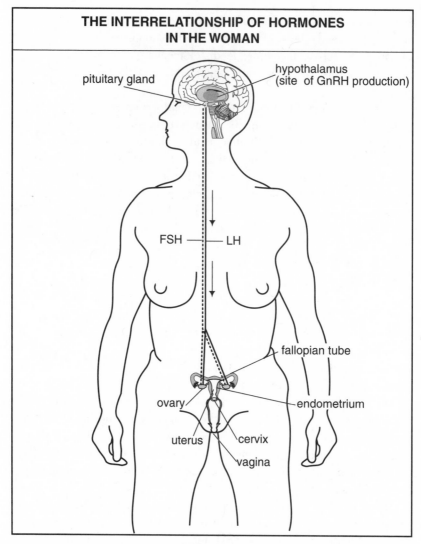

THE INTERRELATIONSHIP OF HORMONES IN THE WOMAN

pituitary gland

hypothalamus
(site of GnRH production)

FSH —— LH

fallopian tube

ovary

endometrium

uterus   cervix

vagina

FIGURE 2-10

example, the hypothalamus may be influenced by stress, pain, environmental changes, diseases in the woman's body, birth control pills, and many forms of medication, including tranquilizers and blood-pressure medication.

The best way to understand this cyclical hormonal process is to trace the woman's hormonal pattern throughout a menstrual cycle. For practical purposes, the menstrual cycle will be considered to begin on the first day of menstruation. The following illustration is based on a 28-day menstrual cycle. However, it is important to remember that many women have somewhat shorter or longer menstrual cycles. In such cases, although the cyclic phases and ovulation occur in the manner described below, their length and timing vary according to the number of days in that particular cycle. For example, a woman with a cycle of 35 days is not likely to ovulate on day 14.

The First Half of the Cycle (the Follicular/Proliferative Phase)

During the first two weeks of the menstrual cycle the body prepares for ovulation (release of one or more eggs from the ovary). During this two-week period until ovulation, the lining of the uterus (endometrium) thickens or proliferates significantly and becomes very glandular under the influence of rising blood estrogen levels. This phase is accordingly often referred to as the *proliferative phase* of the cycle. Because this proliferation of the endometrium occurs at the same time as the development of the ovarian follicle or follicles, it is often also known as the *follicular phase*. From just prior to the beginning of the menstrual period until the middle of that cycle,

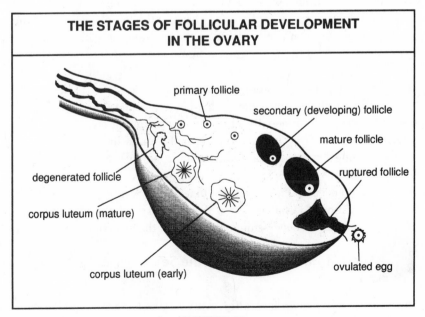

THE STAGES OF FOLLICULAR DEVELOPMENT
IN THE OVARY

primary follicle

secondary (developing) follicle

mature follicle

ruptured follicle

degenerated follicle

corpus luteum (mature)

corpus luteum (early)

ovulated egg

FIGURE 2-11

the pituitary gland releases the gonadotropin FSH in ever-increasing amounts.

The release of FSH causes the formation of follicles in the ovaries and both the production of estrogen and the selection of eggs (usually one per follicle) that are to mature during that cycle. Often as many as 30 or even more follicles begin to develop under the stimulation of FSH from the pituitary gland, but in the natural cycle only one and sometimes two follicles progress to ovulation (see Figure 2-11). The eggs that do not mature ultimately disintegrate and are absorbed into the ovary. This explains why so many eggs are lost during the reproductive life span, although a woman usually ovulates only one, sometimes two, and very rarely three eggs in any particular menstrual cycle.

Responding (usually at the middle of the menstrual cycle) to the rising estrogen levels in the bloodstream, the hypothalamus releases a surge of gonadotropin-releasing hormone (GnRH) when the estrogen reaches a critical level. This rush in GnRH production prompts the pituitary gland to produce a surge of LH, which had been released only in very low, erratic concentrations until this point. It is the sudden surge in LH that actually triggers ovulation.

Ovulation

At ovulation, a muscle connecting the ovary with the end of the fallopian tubes contracts, bringing the fimbriae closer to the follicle containing the egg. The fimbriae then gently massage and vacuum the follicle until the egg, which by this time is protruding from the follicle, is extruded. The fimbriae receive the egg and direct it into a fallopian tube, where contractions transport it toward the uterus.

The follicle collapses once the egg has been extruded and is transformed biochemically and hormonally. It takes on a yellowish color and is then referred to as the *corpus luteum* ("yellow body" in Latin).

All the while that the ovaries are developing eggs and producing hormones, the endometrium is developing in preparation for receiving an embryo. By the time ovulation occurs, it is about three times as thick as it was immediately after menstruation. (In Chapter 8 we will discuss in detail the critical role the endometrium plays in successful implantation of the embryo.)

The Second Half of the Cycle (the Secretory/Luteal Phase)

Once the corpus luteum forms, the ovary begins to secrete the hormone progesterone as well as estrogen, and the levels of progesterone begin to

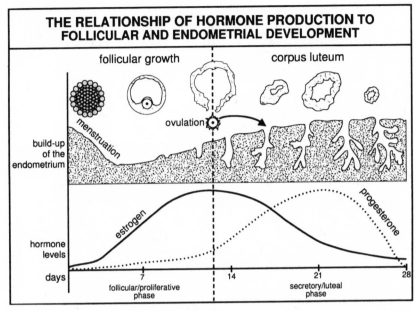

THE RELATIONSHIP OF HORMONE PRODUCTION TO FOLLICULAR AND ENDOMETRIAL DEVELOPMENT

follicular growth   corpus luteum

ovulation

build-up of the endometrium

menstruation

estrogen   progesterone

hormone levels

days   7   14   21   28

follicular/proliferative   secretory/luteal
phase   phase

FIGURE 2-12

rise very rapidly. The progesterone converts the proliferated, glandular endometrium to a juicy structure capable of secreting nutrients that will sustain an embryo—hence the term *secretory phase of the menstrual cycle.* The term *luteal phase* describes the stage during which the corpus luteum produces the progesterone that enhances the secretory environment in the uterus.

The corpus luteum, through the production of both estrogen and progesterone, supports the survival of the secretory endometrium through the second half of the menstrual cycle. The life span of the corpus luteum is about 12 to 14 days if fertilization and implantation of the embryo into the lining of the uterus do not occur. During that period the endometrium is sustained by the estrogen and progesterone produced by the corpus luteum. Once the corpus luteum begins to die, the hormonal support for the lining of the uterus is lost, and two-thirds of the endometrium comes away (often with the unfertilized egg or unimplanted embryo) in the form of *menstruation.* (Figure 2-12 illustrates the relationship of hormone production to follicular and endometrial development throughout the menstrual cycle.)

Should the woman become pregnant, the hormone produced by the implanting embryo and the developing placenta (*human chorionic gonadotropin,* or hCG, which has a similar effect to LH on the corpus luteum) prolongs the survival of the corpus luteum beyond its normal 12- to 14-day life span. The corpus luteum, in turn, continues to produce estrogen and progesterone to maintain the secretory environment of the endometrium, which nurtures the growth of the embryo before it makes contact with the blood system of the mother. Because the corpus luteum continues to exist and produces hormones that nurture the endometrium, the woman will miss her next menstrual period and should then suspect that she is pregnant.

The *placenta* begins to form as the developing embryo establishes a connection with the mother's system. The placenta is both the lifeline between the mother's and baby's blood systems and the factory that nourishes the baby as pregnancy advances. Because the placenta is capable of producing estrogen and progesterone, it soon supplants the need for hormone production by the corpus luteum. The placenta itself supports the endometrium's survival after the 60th to 70th day following the last menstrual period. It has been proved that a pregnancy would continue after the 70th or 80th day even if both ovaries were removed, because the placental hormones themselves are by then fully capable of sustaining the pregnancy.

After ovulation the production of both LH and FSH declines significantly. If pregnancy does not occur, the hypothalamus begins to secrete more gonadotropin-releasing hormone when the corpus luteum begins to die, thus initiating the next menstrual cycle. The same procedure is repeated over and over, with each hormonal cycle setting up the following one, much as each wave in the ocean sets up and determines the character and magnitude of the following wave. It is an indication of nature's ability to maintain biorhythms in a bewildering but organized fashion.

MISCARRIAGE IN EARLY PREGNANCY

Only about one out of every three embryos implants in the uterus long enough to delay the menstrual period. In other words, in two out of every three pregnancies the woman is not even aware that she has conceived.

Even when a pregnancy has been confirmed by a doctor, there is still a 16 to 20% chance of *miscarriage* (expelling of the products of conception after the death of the embryo/fetus) during the first three months of pregnancy. In most cases the reason for this is not apparent. But in those situations where a reason is known, the vast majority of miscarriages are attributed to either a chromosomal abnormality in the developing offspring or hormonal insufficiency.

The use of sophisticated ultrasound techniques to confirm and monitor pregnancies has led to awareness of the phenomenon that not all embryos that implant in the uterus necessarily develop further. (*Ultrasound* is a painless diagnostic procedure that transforms high-frequency sound waves as they travel through body tissue and fluid into images on a TV-like screen. It enables the physician to clearly identify structures within the body and to guide instruments during certain procedures.)

In some cases of confirmed multiple pregnancies, one (or more) of the implanted embryos are absorbed by the body or miscarried and passed through the vagina, thus reducing the number of surviving embryos. This spontaneous reduction in the number of pregnancies appears to be far more common than previously believed, even in those multiple pregnancies occurring without the use of fertility drugs.

An Abnormal Embryo

Early miscarriages usually occur because the embryo is abnormal. Miscarriage is nature's way of protecting the species from an inordinate number of abnormal offspring. These early miscarriages, which mostly occur even before the woman misses her period, are often referred to as *biochemical pregnancies* or *spontaneous menstrual abortions*.

The vast majority of such cases are the consequence of abnormal development in the mixing and replicating of the genetic blueprint. It has been shown that 60% of early pregnancy losses are attributable to a breakdown in the process of early mitosis and meiosis.

As a woman ages, meiosis and mitosis are far less likely to occur without problems, because an older woman's eggs are not able to divide or be fertilized as perfectly as those of a younger woman. This is why the babies of older women are more prone to Down's syndrome and other chromosome abnormalities. (See Chapter 8 for a discussion about how age impacts a woman's ability to have a healthy baby.)

Hormonal Insufficiency

In about 10 to 15% of all pregnancies, the embryo fails to implant because the amounts of hormones produced and the timing of their release were not perfectly synchronized. Such miscarriages, which may occur even if the embryo is perfect in every way, are attributed to *hormonal insufficiency*. This condition is caused by inadequate production of estrogen and/or progesterone during the menstrual cycle.

If hormonal insufficiency occurs because of abnormal hormonal production of estrogen during the follicular phase of the menstrual cycle, it is known as a *follicular phase insufficiency*. If attributable to inadequate production of hormones by the corpus luteum during the second phase of the cycle, it is referred to as a *luteal phase insufficiency*. Miscarriages due to follicular or luteal phase insufficiency may be associated with ovulation that occurs at the wrong time (either too late or too early), the production of inadequate amounts of hormones, an endometrium that responds inappropriately, or a combination of these factors.

Hormonal insufficiency may be perpetuated into early pregnancy, when the embryo is dependent on the survival of the corpus luteum before the placenta develops. Obviously, if implantation is imperfect because of improper hormonal stimulation, then *placentation* (the attachment of the placenta to the uterine wall) might also be defective. Poor placentation might prevent the baby from getting the proper nutrition. As a result, it might grow improperly and could be born too small or too early.

A pregnancy compromised by hormonal insufficiency may delay the onset of the anticipated menstrual period but then result in early miscarriage because of the inadequate hormonal environment. It is sometimes possible, however, to administer certain hormones in early pregnancy to sustain an embryo that otherwise would be lost.

Additional causes of miscarriage include thyroid and other hormonal irregularities, and kidney problems. In addition to causing miscarriage, it is believed that the misuse and abuse of recreational narcotics, psychotropic drugs, and alcohol and nicotine during the first three months of pregnancy, when cell and organ differentiation are taking place, might significantly increase the incidence of birth defects as well as inhibit fetal growth and development.

It is likely that the thickness and quality of the endometrial lining as judged by ultrasound may also be an independent factor that affects the risk of miscarriage. We have come across a number of cases in which there was no apparent cause for recurrent spontaneous abortions (more than three). The only determinant was that the endometrial lining appeared to be thinner than 9 mm in that group.

It is the start in life that counts, and in most cases nature catches its mistakes. The high rate of embryo wastage and early miscarriage when conception occurs naturally may come as a surprise to many couples. However, it should provide a helpful perspective for couples who are considering comparable pregnancy rates offered by IVF and other options.

NATURAL CONCEPTION AND IVF:

TWO PATHWAYS TO PREGNANCY

As we explained in Chapter 2, natural conception occurs when the woman ovulates one or more healthy, mature eggs that unite with the man's normal, healthy, mature sperm. In order for conception to take place, both egg and sperm must travel unimpeded through the reproductive tract and must be fertilized in a supportive (viable) hormonal environment. Until recently, pregnancy was not possible for couples who could not fulfill both requirements: safe transport and a viable environment. In vitro fertilization often solves this problem by bridging anatomical or physiologic disorders that until now have made pregnancy only an elusive dream for many couples (see next section, "Organic and Physiologic Problems That May Prevent Couples from Conceiving Naturally").

CRITERIA FOR NATURAL CONCEPTION

There are five criteria a couple must meet in order for pregnancy to occur naturally:

1. Ovulation of a mature, healthy egg or eggs at the appropriate time, in association with the proper hormonal environment

2. Production of strong, healthy, mature sperm that are deposited in or adjacent to the woman's cervical canal around the time of ovulation

3. A physical-chemical environment that facilitates capacitation (activation) of the sperm as they pass through the woman's reproductive tract

4. A healthy fallopian tube that will promote the passage of sperm and eggs

5. A healthy uterine cavity with no abnormalities that might hinder implantation of the embryo, such as fibroid tumors, polyps that protrude into the cavity of the uterus, or scarring; and an endometrial lining that is thick enough and healthy enough to sustain an appropriate implantation. This can be assessed by ultrasound evaluation of the uterine lining prior to normal ovulation, by hormonal blood testing, and by a biopsy of the endometrial lining just prior to the menstrual period.

Couples who cannot fulfill all five criteria are unlikely to conceive or produce a healthy baby. The following section examines some of the ways in which common disease processes may prevent a couple from becoming pregnant. The remainder of this chapter explains how IVF can compensate for many of these deficiencies and thus enable a heretofore infertile couple to conceive.

ORGANIC AND PHYSIOLOGIC PROBLEMS THAT MAY PREVENT COUPLES FROM CONCEIVING NATURALLY

Neither sex contributes more heavily than the other to infertility problems. Roughly one-third of all infertile couples can trace their infertility to the woman, one-third to the man, and one-third to both partners.

Some Causes of Female Infertility

Organic pelvic disease refers to the presence of structural damage in the woman's pelvis due to trauma, inflammation, tumors, congenital defects, or degenerative disease. The most common cause of infertility in a woman is damaged or blocked fallopian tubes that prevent the egg and sperm from uniting. Sexually transmitted diseases are a major cause of tubal scarring and blockage. In addition, scar tissue that forms after pelvic surgery may also lead to fertility problems.

Conditions such as *endometriosis,* in which the lining of the uterus grows outside the womb (causing scarring, pain, and heavy bleeding), can also damage the fallopian tubes and ovaries. The presence of even a minimal amount of endometriosis in the pelvis is believed to adversely affect fertility by releasing toxic substances that might reduce the potential of the egg to be fertilized. The egg, of course, passes from the ovary through this environment to the fallopian tube, so while the presence of minimal endometriosis might not necessarily adversely affect the transportation mechanism, the egg's exposure to this toxic environment might diminish its ability to become fertilized.

It has also been shown that even the mildest form of endometriosis elicits a local immune response by releasing cells called *macrophages.* These macrophages wander through the pelvis and even into the fallopian tubes destroying the eggs, sperm, and even the embryo. Accordingly, even the most minimal form of endometriosis may reduce fertility by as much as 70% through these mechanisms.

Damaged ovaries may also contribute to infertility. Sometimes an ovary cannot release an egg even though hormonal production is normal and the egg is adequately developed. It is possible for an egg to be trapped within the follicle by scarring or thickening of the ovary's surface. This relatively rare condition may either be hereditary or could be induced by the malfunctioning of structures such as the adrenal gland.

More commonly, diseases such as pelvic inflammatory disease or endometriosis, as well as surgically induced scarring, may anchor the ovaries in an awkward position or form a barrier that prevents the fimbriae from applying themselves properly to the ovaries' surface. Although one or both fallopian tubes may be perfectly free and mobile, the corresponding ovary could be inaccessible and unmovable. In such cases, the egg or eggs would be ovulated into the abdominal cavity instead of being retrieved by the fimbriae.

Abnormal ovulation is another common cause of female infertility. Some women do not ovulate at all, while others ovulate too early or too late in their cycle for a pregnancy to occur and survive. One of the reasons that normal fertility usually wanes after 35 is because ovulation is more likely to become abnormal later in the childbearing years. In addition, it is believed that the quality of eggs decreases as women get older because the eggs' meiotic capacities are diminished by the aging process. The qual of the woman's eggs is one of the major determinants of whether a couple can get pregnant. (We will discuss the relationships to IVF of egg quality, embryo quality, and quality of the endometrium in Chapter 8.)

A woman may also be infertile because disease, surgery, or infection have damaged the lining of her uterus. Damage caused by scarring or the presence of tumors, such as fibroids, may prevent the embryo from attaching to the endometrium and developing properly.

Abnormalities in the size and shape of the uterus can also cause infertility problems. Sometimes women develop an abnormally shaped uterus as a result of exposure to certain drugs their mothers took during pregnancy. A classic example of this disorder is the "T-shaped" uterus and significantly smaller uterine cavity often found in women whose mothers took diethylstilbestrol (DES) during pregnancy.

Some women are unable to produce the cervical mucus that ensures the passage and vitality of the sperm. The production of hostile cervical mucus might be due to infection or abnormal physical and chemical properties in the secretions. Occasionally, surgery or injury to the cervix may have destroyed the glands that produce cervical secretions.

In some cases, women develop antibodies or an allergic response to their partner's sperm. These antibodies may be passed into the cervical secretions and thereby prevent fertilization by destroying or immobilizing the sperm.

Some Causes of Male Infertility

The causes of male infertility are often more difficult to define. Blockage of the sperm ducts is one obvious cause. Generalized blockage may be caused by sexually transmitted diseases. More easily identifiable blockage is caused by a *vasectomy* (voluntary surgery to occlude the sperm ducts for birth-control purposes). While it is usually possible to surgically reconnect the tubes after vasectomy, some men, especially those who underwent the procedure more than 10 years earlier, remain infertile because in the interim their systems have developed an immune reaction that results in the production of antibodies that destroy or immobilize their own sperm.

Another common cause of male infertility is a *varicocele*, a collection of dilated veins around the testicles that hinders sperm function by increasing body temperature in the scrotum. In order for the testicles to produce healthy sperm, the temperature in the scrotum must be lower than it is in the rest of the body.

Ideally, the testicles should have descended into the scrotum shortly after birth, but in some cases they do not reach the scrotum for years. In such circumstances it may be necessary to accomplish this surgically when the man is very young to prevent the testicles from becoming severely damaged, thereby resulting in infertility. In rare cases, abnormal development of the testicles and/or sperm ducts may result from injury, disease, or hereditary abnormalities.

Certain drugs or chemicals in the environment may also inhibit sperm production and function. And as in women, drugs such as DES can also produce abnormalities in the male offspring's reproductive system.

Finally, for reasons that are often not readily apparent, some men lack the adequate hormonal stimulation that is required for proper sperm production.

Unexplained Infertility

For about 10% of all infertile couples, the cause of the infertility cannot be readily determined by conventional diagnostic procedures. Such cases are referred to as "unexplained infertility." Modern IVF technology is making great strides in helping to identify some of the causes of so-called unexplained infertility. Improved testing techniques have made infertility easier to diagnose, and the majority of cases can now be diagnosed and generally are treatable.

For example, recent research has demonstrated that many women with unexplained infertility ultimately are found to have pelvic endometriosis that cannot yet be detected by direct vision during laparoscopy or surgery. For example, a condition called *nonpigmented endometriosis,* in which the endometrium may be growing inside the pelvic cavity with many of the same deleterious effects as overt endometriosis, cannot be detected by direct vision because no visible bleeding has occurred in these lesions. The fertility of these patients may be every bit as much compromised by these conditions as if they had detectable endometriosis.

HOW IVF DIFFERS FROM NATURAL CONCEPTION

This section provides an overview of how IVF adapts the principles of human reproduction to achieve pregnancy. The procedures are described here in general terms and will be discussed in detail in subsequent chapters.

Fertility Drugs Are Used to Produce More Eggs

The administration of fertility drugs promotes the growth of more ovarian follicles than would develop naturally. These drugs also enable more follicles and eggs to mature instead of regressing prior to ovulation. Increasing the number of mature follicles facilitates the retrieval of more eggs and enhances the chance of fertilizing more healthy embryos.

Although an embryo has approximately a 10–15% chance of surviving longer than two weeks in nature, an embryo transferred into the uterus during IVF has no more than an 8 to 10% chance of survival. Because of the IVF embryo's lower odds of surviving, almost all programs transfer several embryos at one time into the uterus in order to give the couple a better opportunity of conceiving.

One category of IVF, however, yields rates of implantation per embryo as high as 20 to 25% (equal to the natural implantation rate). In this category, known as *IVF third-party parenting*, the woman who provides the eggs is not the same woman who will receive the embryos for implantation. In such circumstances, the woman who produces the eggs receives the fertility drugs, her eggs are fertilized, and the recipient's uterus is prepared with hormones to create an ideal lining for implantation. The fact that the recipient of the embryos produces an optimal implantation rate per embryo points towards the importance of the proper endometrial environment.

It may well be that when we stimulate some women with fertility drugs, their ovaries, besides producing the eggs and the estrogen hormone that builds the endometrial lining, also might be producing disadvantageous chemicals and hormones that have an adverse effect on the lining. This problem is obviously sidestepped when the egg donor and the recipient are not the same person. (See Chapter 8 for information about the importance of a healthy endometrium and Chapter 13 for a more detailed discussion about IVF third-party parenting.)

The Chance of a Multiple Pregnancy Is Greater with IVF

Not only is the success rate of an IVF procedure directly related to the number of embryos that are transferred to the woman's uterus, but the more embryos transferred, the more potential fetuses—that is, the greater the risk of twins, triplets, or even larger multiples. (See Chapter 4 for a discussion of the risks of multiple pregnancies and the options available to couples who are confronted with a large multiple pregnancy.)

The risk of multiple babies is not simply a function of the number of embryos transferred but also of embryo quality, which in turn is affected by egg quality. Older women who receive a large number of embryos are far less likely to have multiple pregnancies than younger women receiving the same number of embryos. It is simply a question of embryo viability, which may not be detectable microscopically but from a chromosomal point of view might be an issue.

The impact of age on egg and embryo quality is an unchangeable parameter that must be figured into the multiple-pregnancy equation. Since 1987, we at Pacific Fertility Medical Center in San Francisco have seen only two triplet pregnancies in women undergoing IVF over the age of 40. During that time we studied women over 40 who received eight or more embryos in comparison with women under 35 who received four or fewer. The multiple pregnancy rate was twice as high
red to the women over 40. (We will discuss
ge on egg quality and subsequent embryo

n the Ovaries by Suction

to be ovulated naturally from the follicles,
t of the ovaries through a long needle in a
The needle can be inserted into the follicles
physician monitors its progress on an
6).
ultrasound egg retrievals, it was necessary
r to remove the eggs. In this procedure, a
n as a laparoscope is inserted through an
incision in the navel into the pelvic cavity. The laparoscope enables the surgeon to actually see the pelvic organs and also to aspirate the eggs via a needle inserted through separate puncture sites in the lower abdomen. Laparoscopy is rarely performed for egg retrieval today.

Following retrieval, the eggs are sent to the laboratory for fertilization. Egg retrieval is particularly appropriate when the fallopian tubes cannot retrieve or transport the eggs, when the woman is not able to ovulate properly, or in cases of unexplained infertility.

IVF Bypasses the Fallopian Tubes

The fallopian tubes are entirely bypassed in IVF because the eggs are retrieved directly from the ovaries, and the fertilized embryos are transferred directly into the uterus via the vagina. This is why IVF is particularly suited to women who have damaged or blocked fallopian tubes.

Sperm Are Partially Capacitated in the Laboratory Instead of in the Woman's Reproductive Tract

In vitro fertilization eliminates many of the hurdles that sperm have to overcome, including escaping from the man's semen and passing through the cervical mucus. This is particularly important in cases where the man has an inadequate sperm count or poor sperm function. In vitro fertilization is also helpful in situations when the woman forms cervical mucus that inadequately promotes capacitation or is hostile to the sperm. IVF avoids exposing the egg and embryo to hostile cervical mucus by substituting laboratory procedures for the role of cervical mucus and transferring the embryos directly into the uterus through a catheter.

An IVF Embryo Is Not Likely to React to Either Partner's Antibodies

The body sometimes develops antibodies to sperm after it has become familiar with the spermatic blueprint. Accordingly, as sperm come into contact with bodily immune systems over time, women may build up sperm antibodies and men may even develop antibodies to their own sperm.

In vitro fertilization often evades fertility problems caused by antibodies produced by the man and/or woman. It enables sperm to safely fertilize the eggs in the laboratory without interference from antibodies that would be present in the woman's reproductive tract. The resulting embryos are not affected by those antibodies because mammalian embryos do not have an immunological blueprint. In other words, embryos and fetuses are immunologically inert prior to birth. Thus, the woman's body, which might produce antibodies against sperm, tolerates the embryo because it is an unfamiliar, immunologically inert structure against which her body has not yet developed antibodies.

Abnormalities Are No More Likely to Occur with IVF

An embryo conceived in vitro must be exceptionally healthy and unblemished in order to survive. Keep in mind that an IVF embryo is placed in the uterus a few days earlier in the menstrual cycle than the time it would normally reach the uterus after being fertilized in the fallopian tube. The embryo is also less developed at the time of transfer than it would be when reaching the uterus under normal conditions. In addition, introducing an embryo into an artificially induced hormonal environment (a cycle in which a woman has received fertility drugs) may increase the chances of a luteal phase defect. Therefore, an unhealthy IVF embryo is much less likely to implant than is a healthy one.

Recent studies would indicate that the risk of abnormalities with IVF is no greater than that which occurs in natural conception. In fact, only a handful of major congenital abnormalities have occurred in IVF births reported worldwide. This is so in spite of the fact that women conceiving with IVF are often over 35, when the risk of abnormalities such as Down's syndrome and other chromosomal defects may be as high as 1 in 200 in the population of babies born to women over 35 in the natural setting.

The reason that the risk of abnormalities is no greater than in nature may be that IVF couples are more prudent. Patients who do conceive with IVF are usually well informed and are under close medical scrutiny during early pregnancy. They are more likely than non-IVF patients to present themselves for *tests such as chorionic villus sampling, amniocentesis,* and blood testing for *alphafeto protein* in order to detect the presence of an abnormal fetus at the earliest possible time. This provides an opportunity to perform a therapeutic termination of pregnancy as early as possible.

The same need for an IVF embryo to be especially hardy in order to implant and flourish has resulted in the birth of slightly more girls than boys from IVF. This is because, in addition to nature's slight statistical bias toward girl babies, female embryos and fetuses tend to be hardier than males.

IVF Is Both a Treatment and a Diagnostic Procedure

In vitro fertilization has a built-in diagnostic capability unmatched in nature or by any other method of evaluating or treating infertility. In ideal circumstances there is a 70% or greater chance that any one egg will fertilize in the laboratory. This affords the couple a chance to see whether they are capable of achieving fertilization together. IVF technology has brought to light many instances in which a woman's egg cannot be fertilized by her partner's sperm and sometimes not by any sperm. The reason for this is not always readily identifiable; the problem could lie with the egg, the sperm, or both. If several mature eggs fail to fertilize, this information can help couples make important decisions regarding their future plans. Although the test is not 100% foolproof, failed fertilization should encourage couples to consider micromanipulation, the use of donor eggs, donor sperm, donor embryos, or adoption (see Chapters 13 and 14). No other method of treating infertility enables a physician to reach this diagnostic conclusion. IVF might be called the ultimate fertility test.

Another diagnostic application of IVF would be when, for no readily apparent reason, the fallopian tubes might be unable to properly receive and/or transport the eggs, sperm, and embryos. Because IVF by its very nature bypasses the fallopian tubes, it might—through a process of exclusion—offer both an answer and/or a solution to this problem.

IVF Requires a Heavy Emotional, Physical, and Financial Investment

The most significant difference between IVF and natural pregnancy is that a couple must sacrifice a great deal of their personal privacy before and during the IVF procedure, whereas natural conception is a private matter. An IVF couple must bare some of their deepest secrets and fears to the clinic staff, and allow themselves to be manipulated physically and emotionally as they progress through the procedure. In addition, IVF is inordinately expensive—and there is no second prize if a woman does not conceive following in vitro. The couple will not have another chance at IVF pregnancy without making the same emotional, physical, and financial investment again. In natural conception, there's always next month, and the following month, and hope for the future without the major cost that in vitro fertilization exacts.

CHAPTER

4

IVF STEP 1:

PREPARATION FOR TREATMENT

M ost IVF procedures are based on some variation of the following steps: (1) preparation for treatment, (2) induction of ovulation, (3) egg retrieval, and (4) embryo transfer. All successful IVF programs must be highly organized and exquisitely timed, just as the fertilization process is organized and timed in nature.

Each of these four basic steps in an IVF treatment cycle should be regarded as a hurdle that a couple must overcome before proceeding further. (The term *treatment cycle* refers to the menstrual cycle during which a particular IVF procedure is performed.) Occasionally, a couple may successfully negotiate one hurdle but then be unable to surmount the next step, in which case they would usually begin the treatment cycle anew after allowing the woman's body to rest for a month or two. In general, a couple's chances for successful IVF increase as they put each hurdle behind them.

The descriptions of IVF procedures in Chapters 4 through 7 have been designed to provide an overview of what an infertile couple might expect to experience physically and emotionally during the treatment cycle. Because a truly comprehensive IVF program responds to—and often anticipates—the couple's emotional needs throughout the treatment cycle, some of the techniques that an IVF program might use to address emotional needs are included in the description of clinical procedures. We do not mean to imply that any of these scenarios is the best or the only way that IVF should be performed.

ACCEPTANCE INTO AN IVF PROGRAM

Before being admitted into a hypothetical IVF program, the couple would probably be required to have a complete medical evaluation. They would most likely undergo all the routine steps of an infertility assessment, usually performed by their own primary physician, in order to rule out the possibility that procedures other than IVF might better address their needs. (Chapters 8, 9, and 10 discuss the advisability of exhausting all other options before selecting IVF.)

The couple would probably be required to forward their medical records to the IVF program and are likely to be asked to provide additional background by telephone. They should expect to be encouraged to speak frankly about themselves and their personal habits (including their sexual practices, use and abuse of recreational drugs, general lifestyle, and other parameters that are known to impact fertility). In many programs, including ours, the couple also would be asked to complete detailed psychological inventories. Following a thorough evaluation of the materials submitted by the couple and their primary physician, the medical staff would then decide whether the couple are eligible for IVF.

Once accepted into the program, the couple would probably undergo some orientation, including an explanation of the emotional, physical, and financial commitments that IVF would require. This orientation could take place through letters, other written material, or by telephone. It may also take place on site if the couple are able to visit the clinic prior to commencement of the treatment cycle.

In some programs the couple have to be at the clinic during the entire process, including *induction of ovulation* (usually a series of daily injections). In other programs the couples are encouraged to initiate the induction of ovulation with their own gynecologist, and are required to be on site only for the last few days of the cycle prior to egg retrieval and embryo transfer.

ORGANIZATION OF A TYPICAL IVF PROGRAM

In many programs, one or more nurse-coordinators play an important role in assisting the physician to ensure that the couple receive proper emotional preparation throughout the program. The nurse-coordinators play a central role and administer many treatment procedures that have previously been agreed upon by the entire medical staff.

In such a coordinator-oriented program, the couple could anticipate spending as much if not more time with a nurse-coordinator as with the physician. This is because a nurse-coordinator usually functions as the couple's advocate—the liaison between the couple and all the other members of the IVF team, including the physician. However, this is not meant to imply that both the clinical and administrative roles could not be fulfilled by a physician who has a personality and attitude that will engender a feeling of well-being, relaxation, and optimism. In general, though, nurse-coordinators contribute significantly to the smooth operation of many IVF programs.

It is the responsibility of the person who guides the couple throughout the treatment cycle, whether physician or nurse-coordinator, to explain every step along the way so the couple know exactly what to expect. In addition, the same staff member who is responsible for establishing the initial rapport with the couple should be their contact person throughout their tenure with the program.

In most programs the couple will be introduced to the staff, taken on a tour of the facility, and encouraged to ask a lot of questions. The staff in an IVF program, including the clerical personnel, should be upbeat and encouraging when they deal with infertile couples. The empathic IVF program will provide a relaxing, low-key environment that offers subtle support to both partners during their time of emotional need. Although the couple should be well aware that no program can guarantee a pregnancy, even after several attempts, a congenial atmosphere fostered by the staff should help both partners maintain a mood of guarded optimism.

Some programs, particularly our own, provide access to a counselor with special expertise in the psychological aspects of infertility. Although the participation of a counselor is not essential in order for a couple to conceive, an IVF team member who can predict the way a couple might react, and therefore help improve their tolerance to the emotional roller-coaster ride of IVF, adds another dimension of caring to the program.

TESTS THAT MAY BE CONDUCTED PRIOR TO IVF

Before the couple have a pretreatment consultation with the physician, it is likely that they will be asked to complete some or all of the following tests.

The AIDS Test

A couple should defer pregnancy until they are sure that neither partner carries a disease that can seriously prejudice the health, well-being, and even the survival of the offspring. Although this is a personal decision to be resolved between the man and the woman, the physician enters the picture when IVF is being considered. As the catalyst responsible for creating the circumstances under which a new life might be conceived, the physician has a medical, legal, and moral obligation to make every attempt to ensure that IVF does not lead to the birth of a child who suffers from a life-endangering disease such as AIDS.

Accordingly, many programs require that an AIDS test be done on both partners prior to any IVF procedure. Unfortunately, this test still does not completely rule out the presence of AIDS, because a person may not register positive for up to six months after infection by the AIDS virus. However, the test does provide a good screen to help protect an IVF program from being instrumental in the birth of damaged offspring.

Another reason for administering the AIDS test to all new patients is to protect medical and laboratory personnel who work with the couple.

Some physicians also test both partners for AIDS before the woman undergoes any treatment to enhance fertility, such as reconstructive tubal surgery. This is because if one of the partners is AIDS-positive and a baby with AIDS is born after successful tubal surgery, the couple might argue that they would not have consented to treatment had they known they could transmit AIDS to a baby. We recognize that the decision to undergo AIDS testing is a very personal one and that a physician certainly cannot force anyone to have this test. Nevertheless, we strongly advise that the issue be discussed prior to initiating treatment for infertility.

Some IVF couples have expressed concern that AIDS might be transmitted through fertility drugs because these drugs are derived from the urine of menopausal women. Most experts agree, however, that the AIDS virus does not survive the purification and extraction process to which these drugs are subjected. (See "Cryopreservation as an Option" in Chapter 14 for further discussion about protection against AIDS.)

The Sperm Count

In any IVF program the male partner will almost certainly be asked to submit to a sperm count. The purpose of the sperm count is twofold: (1) to ensure that the sperm's viability and motility are not abnormal and/or have not changed significantly since the last sperm count, which would dramatically affect the couple's rational expectations for successful IVF; and (2) to protect the program from medical-legal liability in case the man has developed an undetected fertility problem since his sperm were last evaluated. The quality of the sperm and the age of the woman undergoing egg retrieval are the two most important factors that enable the physician to predict the likelihood of the couple getting pregnant through IVF.

The Sperm Antibody Tests

More and more programs, including our own, require that the male partner take a sperm test to determine whether he is harboring these antibodies. If a man has sperm antibodies, it could affect the ability of his sperm to fertilize an egg and thereby adversely affect the chances of a successful outcome with IVF. Moreover, the presence of sperm antibodies in the man will significantly influence the manner in which sperm is prepared for the IVF process. In some cases, the presence of high concentrations of sperm antibodies could mandate the performance of *intracytoplasmic sperm injection* (ICSI), where a single sperm is captured in a thin glass needle and is injected into the egg to promote fertilization.

Pelvic Examination

A careful pelvic examination is important in order to evaluate for the presence of irregularities in the contour of the uterus or the adjacent pelvic organs. Their presence might suggest the existence of fibroid tumors, ovarian cysts, swollen fallopian tubes, and other conditions that might affect treatment. In addition, the position of the uterus—i.e., whether it is tipped forward (anteverted), backward (retroverted), or central (axial)—can be assessed; this information will often determine the manner in which the woman will be positioned at the time of embryo transfer.

Uterine Measurement (Mock Embryo Transfer)

At the time of pelvic examination an embryo-transfer catheter should be introduced via the cervix into the uterine cavity. The purpose is to determine the exact depth of the uterus in order to ensure proper placement of the catheter (approximately .5 cm below the roof of the uterine cavity) at the time of the actual embryo transfer. The relative ease or difficulty with which the embryo-transfer catheter is introduced will also assist in planning strategy for the embryo-transfer procedure.

Cervical Cultures and Staining Cervical Secretions for Chlamydia

It is essential to evaluate for the presence of microorganisms that cause pelvic disease such as gonococcus, chlamydia, and *ureaplasma*. Both gonococcus and ureaplasma require specialized culture media. Ureaplasma is a microorganism that occurs in the reproductive tracts of both sexes and might interfere with sperm transport and/or embryo implantation. It commonly produces no symptoms in either partner. If it is present in the cervical secretions it sometimes is also present in the uterine cavity, where it might interfere with implantation. Ideally, the male partner should also be cultured for ureaplasma. If the organism is found in either partner, both should be treated and retested a few weeks after treatment is completed to confirm that the organism has been eradicated.

Hysteroscopy

We routinely perform *hysteroscopy* on women scheduled to undergo IVF if this has not been performed for a year or two. We also would perform hysteroscopy in cases where a *hysterosalpingogram* or the advent of disease/symptoms might suggest the presence of surface lesions in the uterine cavity that have occurred after a prior hysteroscopy was performed. In one out of eight cases, surface lesions that might interfere with implantation are detected by the routine performance of hysteroscopy. This often occurs despite the fact that a recent hysterosalpingogram was reported as being normal and/or that the woman shows no evidence of disease. (See Chapter 9 for more information about these tests.)

The presence of polyps, fibroid tumors that protrude into the uterine cavity, or scarring due to previous uterine infection could significantly reduce the likelihood of success following IVF. Accordingly, they should be treated before the woman undergoes assisted reproduction.

The hysteroscopy can easily be performed under local anesthesia in the doctor's office and does not require any significant postoperative care. We have found this approach to be well received by our patients and are convinced that the routine implementation of hysteroscopy followed by appropriate treatment, when indicated, has prevented numerous women from undergoing otherwise futile attempts at IVF.

Measurement of FSH and Estradiol on the Second or Third Day of a Natural Menstrual Cycle

The measurement of the hormones FSH and estrogen on the second or third day of a menstrual cycle preceding IVF helps evaluate the potential ability of the woman's ovaries to respond to fertility drugs. It also provides information that the physician can use to select the most ideal dosage and regimen of fertility drugs to achieve an optimal response. For this reason, these hormones should be measured in all regularly menstruating women scheduled to undergo IVF.

THE PRETREATMENT CONSULTATION

Once the appropriate tests have been completed, the physician and perhaps the nurse-coordinator will discuss with the couple what to expect throughout the treatment cycle. Although the couple may have a general idea, the physician will reinforce what they have already been told and will encourage them to ask questions.

The physician probably will also outline some of the decisions the couple will have to make in the next few days. These include: (1) how many eggs they wish to have fertilized; (2) what they want to do with any excess eggs; (3) how many embryos they want to have transferred into the woman's uterus; (4) how they wish to dispose of any excess embryos (i.e., through embryo cryopreservation or donation); and (5) how they would deal with a large multiple pregnancy (triplets or larger), should that occur. Although these questions do not all have to be answered at the same time, the physician probably will touch on all of the relevant issues during this consultation in order to give the couple ample time to prepare to make their decisions.

Preparing for the Inevitable Trade-off: Probability of Pregnancy vs. the Risk of Multiple Births

Because an embryo's chance of survival, even in the best of circumstances, is no more than 15% with conventional IVF and only higher (perhaps double) in cases of third-party parenting, enough embryos must be transferred into the uterus to ensure the highest probable birthrate. Some programs transfer a maximum of three or four at a time, others as many as six. It has been our experience that, in women under 40 with a well-prepared endometrium and with both eggs and embryos of good quality, placement of two embryos into the uterus yields a 16 to 20% chance of having a baby, three, about a 25 to 30% probability, and so on up to a maximum of approximately 40 to 45% following the transfer of six embryos.

However, the couple wishing to maximize their chances of pregnancy must be prepared to confront the unavoidable trade-off: the more embryos the greater the risk of multiple births. The multiple-pregnancy rate from IVF in women under 40 is twins in about one out of every four pregnancies, triplets in about one out of every 20, and quadruplets in one out of about every 50 pregnancies.

The physician should explain that it is the number of *viable* embryos transferred to the uterus rather than the absolute number (whether or not they are known to be viable) that determines the risk of multiple gestation. Accordingly, the older patient undergoing IVF will likely be advised to transfer more embryos than would be the case in her younger counterpart. This is because the risk of multiple pregnancies has more to do with the number of viable embryos than with the absolute number transferred. The number of embryos transferred, therefore, should be influenced by the woman's age.

The Risks of a Multiple Pregnancy

The couple must be thoroughly educated on the implications of multiple pregnancy before they decide how many embryos to have transferred. While most women can tolerate a twin pregnancy, a larger multiple pregnancy threatens the well-being of both mother and babies. Moreover, the risks become greater to both mother and babies as the number of fetuses increases.

Risks to the mother that are especially acute during a large multiple pregnancy include high blood pressure, uterine bleeding, and problems associated with a cesarean section (the incidence of cesarean sections increases dramatically in multiple pregnancies).

The primary threat to the physical and intellectual well-being of the babies stems from complications resulting from premature birth. Multiple births often occur prematurely, and the more babies the more premature their birth. Prematurity can cause one or possibly all of the babies to be born brain-damaged and/or with a dangerously low birth weight that can endanger the child's survival.

Selective Reduction of Pregnancy

Because of the serious complications that so often occur in large multiple pregnancies, some IVF programs counsel couples on the concept of selectively reducing the size of a multiple pregnancy as a possible lifesaving measure for the remaining fetuses.

Selective reduction of pregnancy, which is usually performed prior to completion of the third month of gestation, involves the injection of a chemical under guidance by ultrasound directly into one or more developing fetuses. This causes the involved fetus or fetuses to succumb almost immediately, and they are subsequently absorbed by the body. Selective reduction of pregnancy is unlikely to cause a miscarriage in the remaining fetuses—provided that it is done by an expert.

While the possibility of a multiple birth occurring is a good reason for concern, the majority of women who undergo IVF in our program have singleton pregnancies (one baby) in spite of the fact that between four to six embryos are transferred at one time. Experience has demonstrated that the risk of complete miscarriage or damage to the remaining fetuses is very small in cases where the fetuses are not identical, as is virtually always the case with a multiple pregnancy following IVF. This is because IVF multiple pregnancies are almost always fraternal (not identical), and have separate placentas and hence separate blood supplies. We would definitely not advocate performing selective reduction in cases where there are fewer than three babies in the uterus, unless indicated by unusual medical circumstances.

IVF programs might be considered pro-life because IVF by its very nature is the opposite of abortion. Yet many physicians who perform IVF believe strongly that selective termination of pregnancy in the interest of saving life is acceptable or at least presents a possible option. We believe that couples should be made aware, in an unbiased manner, of the option of selective termination of pregnancy in a pro-choice environment.

Constructing a Framework for Decision Making

Before the couple can decide how many eggs should be fertilized, they first have to decide whether they are willing to risk a multiple pregnancy. At this point, the physician might say to them:

There is a difference of opinion as to how many embryos should be transferred into the uterus. What you need to remember is that there is a trade-off. If you put in more embryos you have a higher chance of pregnancy up to a maximum of four. We have found that in women under the age of 35, four embryos is a safe number to transfer to the uterus. In women between 35 and 40 an average of six embryos is safe. And in women over 40, eight or even more embryos can be transferred to the uterus with a relatively low expectation of multiple pregnancies. Remember, it is the number of viable embryos that is important, not the absolute number of embryos we put in. We and others have also observed that freezing leftover embryos in women over 40 is usually relatively nonproductive. These embryos tend not to freeze well, and the survival rate after thawing is very low.

How many embryos are you going to want? Because we can usually fertilize between 70% of the eggs that we retrieve, we would encourage you to have at least six or seven eggs fertilized if you want four embryos transferred. If you want four embryos but fertilize only four eggs, you may only end up with no more than two or three embryos. That's the chance you take.

The couple who want four embryos to be transferred thus have two choices: (1) to fertilize only four eggs to preclude the development of more than four embryos (but possibly ending up with only two or three) or (2) to fertilize more than four eggs in order to have four healthy embryos.

If the couple are willing to risk a multiple pregnancy, the physician may then ask whether they would consider selective pregnancy reduction should it become necessary. If they agree to the concept of selective pregnancy reduction, the physician might be satisfied to subsequently transfer four embryos. Some couples, for moral, ethical, or religious reasons (or simply because they want to enhance their chance of conceiving), want every embryo transferred but refuse to consider selective pregnancy reduction; that is their choice. In such cases, the physician is likely to agree to transfer as many embryos as the couple request but could be expected to ask them to sign a release.

This is a particularly crucial discussion because it leads to related decisions. For example, what if all the eggs fertilize, resulting in eight or nine embryos? If four are transferred, what should be done with the remainder? Should they be frozen for IVF use during a subsequent menstrual cycle? Should they be donated to another couple? Might they be used for research? Or should they be discarded? The same questions apply to eggs that the couple choose not to have fertilized.

Decision: How Many Eggs Should Be Fertilized?

The couple must decide prior to the egg retrieval how many eggs to fertilize and what should be done with any excess. They can delay deciding how many embryos to transfer, how they wish to dispose of any extra embryos, and how they intend to deal with a large multiple pregnancy—but only until immediately before the embryo transfer. They would be well advised to thoroughly discuss their feelings and preferences in the interim because all of these decisions are fundamentally interrelated.

CHAPTER

5

IVF STEP 2:

INDUCTION OF OVULATION

A woman undergoing IVF is given fertility drugs for two reasons: (1) to enhance the growth and development of her ovarian follicles in order to produce as many healthy eggs as possible and (2) to control the timing of ovulation so the eggs can be surgically retrieved before they are ovulated. In vitro fertilization is performed relatively infrequently in natural cycles (cycles in which no fertility drugs are administered) because in such cases it is unlikely that more than one or two eggs can be retrieved at a time.

The ovulation of more than one egg that has been induced through the administration of fertility drugs is known as *superovulation*. The term *controlled ovarian hyperstimulation* (COH) encompasses the concept of deliberate induction of superovulation but also refers to production of an exaggerated hormonal response that favors implantation of the embryo into the endometrium. The terms stimulation, superovulation, and controlled ovarian hyperstimulation are often used interchangeably. However, we are using specific terms such as these to convey slightly different emphases.

The maturation and health of the eggs developing in the ovaries are measured by the concentration of the hormone estrogen (the *estradiol,* or E_2 level) in her blood and/or visualization of the developing ovarian follicles by ultrasound. (Estradiol is a hormone secreted into the bloodstream by the growing ovarian follicles.) It has been shown that the greater the degree of controlled ovarian hyperstimulation, the more eggs will be available for retrieval.

Ovulation can be expected to occur about 38 to 42 hours after a woman is optimally stimulated. Egg retrieval therefore has to be scheduled for a few hours prior to the anticipated time of ovulation so the eggs can be retrieved *before* being expelled into the abdominal cavity. As will be explained later, the methods of assessing the degree of stimulation, predicting the time of ovulation, or inducing ovulation vary according to the fertility drug used.

Induction of ovulation makes physical and emotional demands on the couple. The woman should expect to undergo daily administration of a fertility drug, usually by injection, as well as blood tests and/or ultrasound evaluations to monitor her progress. In addition, both partners should be prepared to cope with the strong emotions that some women experience because of the hormonal changes introduced by the fertility drugs. As one nurse-coordinator explained:

It doesn't take anything to make women emotional at this point. I have at least one patient a week sobbing in my office just looking at the plants. Sometimes they will even cry over dog-food commercials. The couple have to be prepared for this, and the man should be especially supportive and tolerant.

AN INDIVIDUALIZED APPROACH TO INDUCTION OF OVULATION

The hormonal dosage is tailored to each woman's unique circumstances. The protocol is defined by the woman's age, her FSH and blood estradiol levels on the second or third day of a preceding natural menstrual cycle, and her previous response to fertility drugs if she has undergone stimulation in the past.

FERTILITY DRUG THERAPIES

The following section outlines the most common fertility drugs used in the United States. In general, the woman's response to these drugs will depend on her pattern of ovulation and her age.

Clomiphene Citrate

Clomiphene citrate is the most popular method of inducing ovulation and in the 1980s was the most widely used method for COH in preparation for IVF. *Clomiphene citrate* is a synthetic hormone that deceives the hypothalamus into thinking that the body's estrogen level is too low. In response, the hypothalamus releases GnRH (gonadotropin-releasing hormone), which in turn prompts the pituitary gland to release an exaggerated amount of FSH (follicle-stimulating hormone). As happens in nature, the increased secretion of FSH stimulates development of the follicles, ultimately resulting in ovulation. The growing follicles secrete estrogen into the bloodstream, thus closing the feedback circle that the hypothalamus initiated in response to the anti-estrogen properties of clomiphene. (When marketed in the United States, clomiphene is also known as Clomid and Serophene.)

Administration of clomiphene citrate enhances the normal cyclical pattern of follicular development and ovulation. If initiated as early as day 2 or day 3 of the menstrual cycle, it usually induces ovulation on day 13 or 14 of a regular 28-day cycle. If administered later, such as on day 5, ovulation could occur as late as day 16 or 17, and the length of the cycle may be extended. If the woman does not stimulate appropriately on the original dosage of clomiphene, the dosage may be increased to achieve optimal stimulation. We sometimes administer hCG to the patient once ultrasound examinations and hormonal evaluations confirm optimal follicular development. In such cases ovulation will usually occur about 38 hours later. (See "The Second Half of the Cycle [the Secretory/Luteal Phase]" in Chapter 2 for a review of the role of hCG.)

Finally, the prolonged usage of clomiphene for more than three consecutive cycles may lead to the accumulation of one of its components (zuclomiphene), which will reduce the amount and alter the quality of the cervical mucus, with negative implications for the passage and capacitation of the sperm. It is also likely to thin the uterine lining and thereby reduce the chances of a healthy implantation. This is the reason why the prolonged usage of clomiphene without at least one month's break every three months is associated with reduced pregnancy rates and a much higher rate of spontaneous abortion, should pregnancy occur. It also explains why 80% of viable pregnancies that occur following the use of clomiphene are conceived during the first three months of stimulation and why hardly any pregnancies occur at all in women who have taken clomiphene more than five or six months without a break. One month's hiatus is sufficient to allow for the elimination of zuclomiphene and will restore the potential to respond optimally to clomiphene.

It has been observed that few women over 40 respond well to clomiphene. In spite of the fact that they appear to ovulate on clomiphene treatment, they frequently develop poor mucus and a poor endometrial lining from the inception of clomiphene administration. We accordingly believe that clomiphene treatment is relatively contraindicated in women over the age of 40.

Two major advantages of clomiphene are its relatively low cost and the fact that it can be taken orally instead of by injection. A distinct disadvantage is that when administered alone it does not stimulate the growth and maturation of as many follicles as do alternative therapies such as *human menopausal gonadotropin* (hMG) or clomiphene plus hMG; accordingly, fewer eggs can be retrieved. (When marketed in the United States, hMG is also known as Pergonal. See "Human Menopausal Gonadotropin [hMG]" later in this chapter.)

Side Effects of Clomiphene

The side effects associated with clomiphene are related to the follicular development the drug has stimulated. When administered alone, a luteal-phase defect may result if the follicles do not develop properly. This would hinder implantation by preventing the endometrium from responding optimally to the progesterone produced by the corpus luteum. Clomiphene may also interfere with the nurturing effect estrogen must have on the developing endometrium. In addition, traces of clomiphene that might linger in the woman's circulatory system for many weeks may inhibit the normal function of enzymes produced by the developing follicular cells.

Too high a dose of clomiphene may cause follicles to grow too rapidly, producing large fluid-filled collections known as cysts. This may lead to tenderness and swelling of the ovaries, visual disturbances, and hot flashes similar to those at menopause.

Finally, too high a dose of clomiphene may decrease the amount of cervical mucus produced and may also reduce its quality, with negative implications for the passage and capacitation of the sperm. Too high a dosage may also decrease the thickness of the uterine lining.

Safety of Clomiphene

Some studies have suggested that clomiphene citrate has caused birth defects or a higher miscarriage rate in laboratory animals and could, therefore, potentially threaten human offspring. We, however, believe that when clomiphene is taken under proper supervision these risks should not be of major significance.

The fear that clomiphene might cause birth defects arises from the fact that its inner structure, or nucleus, is very similar to that of the hormone DES, which is known to have caused so many birth defects when administered to pregnant women. Although it is theoretically possible that clomiphene might cause such defects, birth statistics do not indicate an increased birth-defect rate after stimulation with the drug. The laboratory studies mentioned above should not be ignored, however, but should be heeded as a guide to safe, prudent administration of fertility drugs.

We caution that clomiphene citrate should be taken only when it is absolutely certain that the woman is not pregnant. (The appearance of a menstrual period does not provide adequate certainty because more than 10% of women might bleed during early pregnancy. Assessment by a physician, or even a home pregnancy test, provide greater assurance that a pregnancy does not exist.)

The administration of clomiphene as a fertility agent over a series of months might also promote ovulatory problems. It has been observed that in one out of five cases where clomiphene is administered, the egg remains trapped in the follicle after ovulation. Therefore, the practice of physicians saying to patients, "Here's some clomiphene—take some each month and call me if you miss your period" should be deplored.

But if clomiphene citrate is taken under proper supervision and the woman has previously determined that she is not pregnant, its safety is beyond question. This has prompted many IVF programs to continue using clomiphene. Those that do so, however, invariably report a lower pregnancy rate than that which can be achieved by other methods of controlled ovarian hyperstimulation.

Human Menopausal Gonadotropin (hMG)

The fertility drug hMG (Pergonal) contains equal amounts of the gonadotropins FSH and LH. This drug is derived from the urine of menopausal women, which is a good source of both FSH and LH. This is because a menopausal woman's pituitary gland, in response to a feedback message that her ovaries are no longer producing enough estrogen, increases the output of FSH and LH in an effort to restimulate the failing ovaries. The excess FSH and LH is excreted in the urine. Urine used for hMG is distilled, filtered, and purified by an expensive process. At the time this book is being written, one ampule of hMG costs about $50 to $60 in the United States, and the average woman might require 25 or more ampules per treatment cycle.

Human menopausal gonadotropin, which is used in many successful IVF programs, is considered to have many advantages. Instead of influencing the hypothalamus and pituitary gland to produce more hormones to stimulate follicular development (as is the case with clomiphene), hMG acts directly on the ovaries. In addition, it does not inhibit the function of estrogen or the enzymes of the cells lining the follicles.

If administered in sufficient amounts beginning early enough in the menstrual cycle, hMG will prompt the maturation of a large number of follicles. Although the average number of eggs usually retrieved from a woman younger than 40 after hMG stimulation—provided she has two ovaries—is between six and 10, retrievals of more than 50 eggs have been reported.

Because hMG cannot be absorbed through the stomach into the bloodstream, it must be administered by injection rather than in pill form. The usual injection schedule is from day 2 or 3 through day 8 to 12 of the menstrual cycle.

One of the most significant attributes of hMG is its safety-valve effect on ovulation. No matter how well stimulated a woman becomes when she takes hMG, she will be unlikely to ovulate until she receives an injection of hCG. Thus, if for any reason it is determined that the woman should not progress to ovulation, the hCG is simply not administered.

Side Effects of hMG

Many women taking hMG report breast tenderness, backaches, headaches, insomnia, bloating, and increased vaginal discharge, which are directly due to increased mucus production by the cervix.

Luteal-phase defects (inadequate production of progesterone by the corpus luteum to sustain the endometrium) are also known to occur in association with hMG therapy. However, endometrial biopsies have shown that the development of the uterine lining of patients stimulated with hMG is usually a few days ahead of that which could be expected in unstimulated cycles. Thus, hMG helps to synchronize development of the endometrium with growth of the follicles and eggs. This synchronization is a critical prerequisite for successful implantation because IVF embryos are usually transferred to the uterus a few days earlier than they would reach it under natural circumstances. Therefore, accelerated endometrial development enhances the chances that the young embryos will implant after their transfer to the uterus.

Possible side effects of hMG overstimulation that threaten the woman's well-being include enlargement and "weeping" of the ovaries, a condition in which a large amount of fluid is exuded into the abdominal cavity. In severe cases this can cause the abdomen to distend severely and may even compromise breathing. In rare cases the kidneys or liver may fail and the woman may stop producing urine, which can be life-threatening. In very severe cases her blood may lose its ability to clot properly. These situations, however, are extremely rare and are sometimes caused by inappropriate use of hMG. They are highly unlikely to occur in the properly managed cycle.

It is significant that hMG is unlikely to produce any serious persistent side effects until the woman receives the injection of hCG to stimulate ovulation. Thus, the physician has ample time to assess her status and withhold the hCG if it appears that she might develop major side effects. (Such an assessment is made on the basis of blood estradiol values or ultrasound examinations immediately prior to administration of hCG.) This built-in protective advantage shields almost all women being treated with hMG (administered either alone or in combination with clomiphene) from the serious hazards of overstimulation (see "The Injection of hCG—A Safety Valve" later in this chapter).

However, in certain circumstances inadvertent ovarian hyperstimulation might occur. It has been demonstrated that if more than 30 follicles develop following stimulation with fertility drugs and the woman's plasma estradiol at its highest level exceeds 6,000 picograms per milliliter, there could be as much as an 80% risk of the life-endangering complications described above. Until recently, the only way to prevent these complications from occurring was by withholding hCG in those cases where inadvertent overstimulation appeared to be taking place.

We recently introduced a new approach that permits the cycle of treatment to continue while eliminating any significant risk of severe ovarian hyperstimulation. This new method can only be applied in cases where patients are being concurrently treated with GnRHa (the gonadotropin-releasing hormone agonist). In cases where the advent of severe hyperstimulation is suspected, the hMG therapy is withheld while the GnRH agonist treatment is continued, and the woman undergoes daily blood estradiol measurements until the concentration drops to a safe level. At that time hCG is administered, regardless of the number of developed follicles or the number of eggs retrieved; thus, these women do not develop life-endangering complications. Although the fertilization rate of the eggs appears to be reduced using this method, the pregnancy rate does not appear to be compromised. We have termed this method *prolonged coasting*. Prolonged coasting prevents canceled cycles and with them, canceled dreams.

Finally, the entire contents of the follicles are removed during egg retrieval, thus reducing the likelihood that the ovary will "weep." For this reason, serious side effects from hMG are much less likely to occur in women who undergo egg retrieval than in those who do not.

Variations in Response to hMG

Some women stimulate well after relatively small doses of hMG. Others require two, three, or even four times that dosage to achieve the same effect. In the past, selecting the proper dosage was a trial-and-error process. There was simply no way to predict how a particular woman might respond. Each woman is unique, and each can be expected to react differently to hMG. However, about 80% of women respond appropriately to an average injection.

We now measure FSH (follicle-stimulating hormone) and estradiol (E_2) in the woman's blood on the second or third day of a natural menstrual cycle preceding the IVF cycle. We use the levels of these hormones to predict the probable way she will respond to a variety of stimulation methods. We believe these tests are also valuable in selecting the most appropriate dosage and regimen of fertility drugs to be administered.

Despite these refinements, however, stimulation for IVF is still somewhat of a hit-or-miss procedure. For example, when a woman has used up most of her lifetime egg budget and is left with less than a critical number of eggs, she begins to enter a phase of hormonal change known as the *climacteric*. The climacteric is associated with a loss of fertility, the onset of hot flashes, and mood changes. It ultimately culminates with the total cessation of menstruation between the ages of 40 to 55, a process called the *menopause*. The ovaries still produce hormones after menopause, but they are released in a constant rather than cyclical manner.

When a woman fails to become stimulated on the first try, hormone tests are indicated to ensure that she is not in the climacteric as well as to determine if hormonal abnormalities or other conditions might be inhibiting her sensitivity to hMG. If she is not in the climacteric and no other abnormalities are detected, then it can be anticipated that she will eventually respond to an adjusted dosage of hMG. She can begin another round of hMG therapy with an adjusted dosage after she lets her body recover for a month or two.

Follicle growth and development, egg maturation, the number of eggs that can be retrieved, and the risk of side effects are directly related to the patient's response as evaluated by blood-estrogen levels and/or ultrasound, not to the dosage of hMG. Therefore, it is illogical to fear administering an escalating dose of hMG after a poor response to a standard dosage. What is important is to monitor the *individual's* response to the drug.

The tremendous interpersonal and intrapersonal variations in response to hMG might be attributable to one or both of the following factors: First, it could be that hormonal and biochemical factors governing the response to hMG vary in different women or even at different stages of the same woman's life. For example, as we noted earlier in this chapter, age influences the woman's ovarian receptivity to hMG, especially as she nears the climacteric. In addition, some women simply will not respond consistently to administration of the same dosage of hMG from month to month.

A second possible cause of variations in response is that separate batches of hMG might have differing biopotencies. Gonadotropins such as FSH consist of a combination of many similarly structured components called *isohormones*. Unfortunately, not all these isohormones are biologically equivalent in their activity. Thus, the activity of the isohormones may vary from batch to batch of hMG and related drugs. Fortunately, much research is currently under way to identify the active ingredients of gonadotropins and thereby better standardize their biopotency.

Combination of Clomiphene and hMG

A few IVF programs today still administer a mixture of clomiphene and hMG for controlled ovarian hyperstimulation. While the vast majority of IVF programs in the United States no longer use this combination, a significant number of pregnancies have been reported by the few programs that still cling to this method. One of the reasons for using this combination is because clomiphene increases the ovaries' sensitivity to hMG, thereby reducing the dosage of hMG that must be administered. Thus, the overall cost of the fertility drugs is significantly decreased by reducing the required amount of expensive hMG. A second reason for administering these drugs in combination is to simplify their administration (clomiphene can be taken in pill form although hMG must be injected).

However, we believe that the administration of this combination has several drawbacks. First, because clomiphene has the ability to induce ovulation, a combination of clomiphene and hMG may cause spontaneous ovulation even without the administration of hCG. In such a case, egg retrieval might inadvertently take place after ovulation has occurred, resulting in fewer eggs being retrieved. In addition, some physicians feel that the combination of the two drugs makes it difficult to pinpoint the amounts of each that should be changed when the overall dosage must be adjusted in subsequent treatment cycles.

Side effects associated with the use of hMG and related to the degree of stimulation as measured by estrogen levels also apply to the use of clomiphene plus hMG. Accordingly, proper management of the treatment cycle should limit the risks of side effects from the clomiphene–hMG combination.

Purified FSH

Purified FSH is derived by processing and purifying hMG to eliminate most, if not all, of the LH. It has been postulated that the LH component of hMG directly stimulates the tissue surrounding the ovarian follicles (*ovarian stroma*), which propagates precursors that in turn produce male hormones (*androgens*). Some of these androgens may filter into the surrounding follicles and adversely affect egg development. It is also possible that some of the androgens could inhibit the proper development of the endometrial lining. In this way, ovarian androgen production induced by LH could have a deleterious effect on egg and embryo quality as well as on the potential for healthy implantation in the endometrium. Accordingly, we have largely substituted FSH for hMG in most cases of ovarian stimulation in preparation for IVF. Finally, as with hMG, variations in both interpersonal and intrapersonal responses due to the influence of isohormones are known to occur with purified FSH.

Purified FSH also has advantages in treating selected infertility problems such as ovaries with multiple small cysts (*polycystic ovarian disease*), but this situation rarely applies to the IVF setting.

Recombinant FSH

Purified FSH derived from hMG processing is not completely free of LH. Fortunately, a new technology has been developed that makes it possible to produce purified FSH by bacteria through genetic engineering. Known as recombinant FSH, it appears to be more bioactive than purified FSH derived from the processing of hMG.

Gonadotropin-releasing Hormone Agonist (GnRHa)

In about 25% of cases where women receive hMG and FSH alone or in combination, a premature release of the hormone LH causes the follicle to stop developing and damages the eggs. This results in a fall in the plasma estradiol concentration and mandates cancellation of the treatment cycle. Moreover, studies show that many of the interpersonal variations in response to fertility drugs occur because some women release higher levels of LH into their blood than others. This finding has led to the suggestion that substances able to inhibit the release of LH in a woman being treated with fertility drugs may help standardize responses to the drugs, both among individuals and from cycle to cycle for the same woman. These agonists can be taken intranasally as snuff or a spray, or by daily injection.

Drugs that inhibit the release of LH, called GnRH (gonadotropin-releasing hormone) agonists, are now commonly used prior to or in combination with hMG and FSH. This treatment reduces the number of canceled cycles by more than 60%. (GnRHa is marketed in the United States as Lupron, Synarel, and Nafarelin.)

One of the minor disadvantages of using GnRH agonists is that they tend to reduce the ovarian sensitivity to hMG and FSH, thus requiring higher dosages of these drugs as well as their administration for a longer period of time than would otherwise have been necessary. However, the advantages of combining these two agents far outweigh the disadvantages.

The use of GnRH agonists represents a significant breakthrough in the treatment of infertility.

EVALUATION OF FOLLICULAR DEVELOPMENT

In order to properly schedule surgical retrieval of eggs either by ultrasound needle-aspiration or by laparoscopy, the IVF physician must be sure that proper follicular development has occurred. In vitro fertilization programs rely heavily on the woman's daily blood-estrogen levels and daily ultrasound measurements of follicular dimensions in order to fine-tune this scheduling. Usually by day 9 to 14 of the cycle, follicular development is evaluated and a go/no-go decision can be made about whether to proceed to egg retrieval.

The Role of Vaginal Ultrasound Evaluation

Ultrasound is a painless procedure that transforms high-frequency sound waves as they travel through body tissue and fluid into images on a TV-like screen. The woman can feel the pressure of the ultrasound transducer in her vagina, but she cannot feel or hear the sound waves. There is no evidence that the ultrasound waves cause any damage. Ultrasound enables the physician to see the woman's ovaries clearly and to identify, count, and even measure the fluid-filled follicles as they develop. As ovulation approaches the follicles tend to get larger, and monitoring their development by ultrasound provides an important indicator of the expected time of ovulation.

The Importance of Measuring Plasma Estradiol and Performing Ultrasound Follicular Assessments

A traditional method of monitoring the response to hMG/FSH is by measuring the level of the hormone estrogen (the estradiol/E_2). The estradiol level gives an approximate indication of how many eggs the physician might expect to retrieve: in general, the higher the estradiol level, the more eggs. This test will usually be done daily during the latter part of the treatment cycle preceding egg retrieval. In some programs the estradiol level is measured throughout the treatment cycle.

When hMG/FSH is administered, follicular size as measured by ultrasound may have little bearing on the likelihood that a follicle will produce a healthy egg. Even the smallest mature follicles, under pure hMG stimulation, commonly produce healthy, mature eggs.

If a woman taking hMG/FSH does not stimulate high enough after a week or so, she can continue hMG medication for a few more days while her response is monitored by blood tests and/or ultrasound examinations. Such a delay should not significantly decrease her chance of getting pregnant. In certain cases, a woman who has not received GnRHa will show an optimal response to hMG/FSH followed by an unexpected drop in hormone levels prior to egg retrieval. This is usually an indication that the hormone LH has been spontaneously released prematurely, thus threatening the health of the follicles and eggs. When this occurs the physician should consider canceling the treatment cycle, and should reassess both the method and dosage of stimulation in a subsequent cycle.

The Injection of hCG—A Safety Valve

When blood estradiol levels and/or ultrasound assessment indicate that the follicular development of a woman taking hMG is enough to produce an adequate number of eggs but unlikely to cause dangerous side effects, she will be given an injection of hCG (human chorionic gonadotropin) to ripen the follicles and eggs for ovulation. The hCG triggers ovulation, which occurs within 38 to 42 hours, in the same manner as does the surge of LH in nature.

Similar in structure to LH, hCG is favored for the induction of ovulation. Human chorionic gonadotropin, the same hormone that is measured to assess whether a woman is pregnant, is derived from the urine of pregnant women. Because hCG is broken down and made inactive when it passes through the stomach if taken in pill form, it must be injected in order to be transported directly to the ovaries.

After the administration of hCG, egg retrieval is scheduled to be performed within about 36 hours, i.e., prior to the anticipated time of ovulation. The follicles will then continue to grow until the eggs are retrieved or ovulation occurs.

Because hMG/FSH is unlikely to produce any serious side effects until the woman receives the injection of hCG, the decision to withhold hCG provides additional protection against the hazards of overstimulation. However, the newfound ability to prevent the complications of severe ovarian hyperstimulation syndrome through "prolonged coasting" has essentially removed most of the major risks associated with the use of fertility drugs.

Does the Use of hMG and FSH Increase the Risk of Ovarian Cancer?

A paper that appeared in the medical literature in January 1993 suggested that the use of fertility agents might increase a woman's risk of subsequently developing ovarian cancer. The authors of the paper, a group from Stanford University, computer-analyzed data from 12 studies on ovarian cancer patients done in the late seventies and early eighties. They compared data about ovarian cancer patients with data about women of similar age and background who did not have the disease, looking for patterns relating to the incidence of ovarian cancer in infertile women who received fertility drugs.

In our opinion, this report makes unsubstantiated deductions, for the following reasons. The only way to determine whether there is a link between fertility drugs and an increased occurrence of ovarian cancer would be through a prospective controlled study (a study in which data will be collected in the future) that compares two groups of infertile women: those who receive fertility drugs and those who do not.

Until the widespread introduction of IVF in the late eighties, the reason most women were treated with fertility drugs such as clomiphene and hMG was because of abnormal or absent ovulation. More recently, due to the advent of assisted reproduction and intrauterine insemination, most women who receive fertility drugs ovulate normally; problems other than dysfunctional ovulation are at the root of their infertility.

Since this study reported on the use of fertility drugs prior to their widespread administration to normally ovulating women (for assisted reproduction), it is clear that this study was biased. It would obviously take a very large controlled study to establish a definitive link between ovarian cancer and the use of fertility drugs. It is interesting that a few years ago the same journal that reported a link between ovarian cancer and the use of fertility drugs reported a follow-up on a large group of infertile women who received fertility drugs over a period of 12 years. This study failed to show any link with ovarian cancer.

Perhaps most important, the study reported in January 1993 did not adequately address the fact that because pregnancy has a protective effect against the development of ovarian cancer, those women who conceive through the use of fertility drugs might well have benefited from this treatment. The ultimate answer would come from a large prospective study aimed at evaluating the incidence of ovarian cancer in women who conceive following the use of fertility drugs versus those who fail to conceive in spite of such treatment. Such a study would take decades to complete but would certainly identify the risk-benefit ratio associated with the use of fertility drugs.

We believe that the report suggesting an increased incidence of ovarian cancer with fertility drugs was poorly conceived, untimely, and inappropriate. It has added to the hurt of couples already ravaged by the emotional roller coaster of infertility.

Moving On to Egg Retrieval

The woman who is optimally stimulated will, in our opinion, usually demonstrate a continuing rise or at least maintain a sustained level of estradiol for at least one day upon discontinuation of hMG, FSH, or clomiphene plus hMG. This would confirm that the follicles and eggs are continuing to develop toward optimal maturation. Moreover, it has been demonstrated that a large drop in the estradiol level after hMG is discontinued will often be associated with a decline in the quality of the eggs to be retrieved. Accordingly, many programs allow a day to elapse after discontinuing hMG/FSH therapy before administering hCG in order to assess whether the estradiol level will continue to rise. If, however, it has been sustained or shows a progressive rise, the patient is eligible for egg retrieval.

The timing of egg retrieval depends on the stimulation technique used. When clomiphene is administered alone, egg retrieval is often scheduled for 27 to 28 hours after the onset of the LH surge is detected in the woman's blood. If she has been stimulated with hMG, the optimum time for egg retrieval is about 36 hours after the final hMG injection is administered.

The average number of eggs retrieved varies from program to program, depending on the method and the degree of stimulation used (the higher the hormonal levels, the more eggs can be expected). We average between six and 10 eggs per retrieval attempt, although we have retrieved as many as 50 eggs at one time.

Overcoming the induction-of-ovulation hurdle is particularly significant in IVF. But in IVF, as in other important events in life, things don't always go as planned. Therefore, a reputable IVF program should counsel couples in preparation for the possibility that they may experience a poor stimulation cycle.

However, couples able to negotiate the induction-of-ovulation hurdle have a right to be guardedly optimistic about their overall chances of success. This is how one IVF physician encourages his patients and at the same time helps them maintain realistic expectations:

While the level of hormones and the ultrasound findings roughly correlate with the chances of retrieving a large number of eggs, this doesn't always hold true. Sometimes the follicles don't want to give up the eggs, or scar tissue may prevent us from reaching the ovary. And just because we retrieve an egg doesn't mean it will fertilize.

If we get a lot of eggs, that's great. But if we don't, I always emphasize that we have had pregnancies result from the transfer of just one embryo.

CHAPTER

6

IVF STEP 3:

EGG RETRIEVAL

Arrival at Step 3 represents a major accomplishment for IVF candidates because it means that the woman has been optimally prepared, both physically and emotionally, for egg retrieval. Now, for the first time in the treatment cycle, she and her partner have a realistic expectation of conceiving, since the IVF pregnancy rate is usually based on the chance of getting pregnant after undergoing egg retrieval. Their chance of success is the pregnancy rate quoted by the program they have selected.

The egg retrieval phase exacts the greatest physical, emotional, and financial investment the couple will be expected to make in the entire treatment cycle. From egg retrieval onwards, the financial investment in IVF escalates sharply by the hour, largely because of the costs involved in the egg retrieval procedure, laboratory fees for fertilization, and the embryo transfer. This outlay is particularly burdensome in the United States, where most couples must assume the entire expense since few insurance companies will fund these procedures.

THE PRE–EGG-RETRIEVAL CONSULTATION

Prior to egg retrieval, the couple should have a refresher consultation with the physician who will actually perform the procedure. They should also meet with the rest of the IVF team that will be involved, including the nurses who will provide postoperative care.

The physician should explain the procedure in detail and describe how the woman might expect to feel afterwards. In addition, the physician should point out that although serious complications are highly unlikely, no one should undergo any kind of surgical procedure with the idea that it is devoid of risk. Although extremely rare, complications may include infection, bleeding, and injury to surrounding structures such as the bowel, bladder, or major blood vessels.

Because women who receive fertility drugs very often have corpus-luteum insufficiency, from the day of egg retrieval onwards many programs administer injections or vaginal suppositories containing progesterone to augment the production of progesterone by the corpus luteum (see Chapter 2, "Ovulation"). If this is the case at that particular clinic, the physician probably will discuss it with the couple at this point.

A consultation should also take place with the couple's anesthesiologist. The anesthesiologist should also review the woman's medical history, looking for conditions that could complicate the procedure. In the event that any such factors are detected, the anesthesiologist may call for an electrocardiogram, blood or urine tests, or other appropriate diagnostic measures in order to ensure that the surgery can proceed safely.

Finally, prior to the administration of any medication, the woman and her partner should be asked to decide what to do with the eggs that are retrieved. The physician should reiterate the various scenarios previously discussed during the initial consultation, including a reminder that the more embryos transferred the higher the pregnancy rate—with the concurrent risk of multiple pregnancy (see Chapter 4). The couple will usually be expected to complete and sign a directive stating how many eggs they want fertilized, how many embryos (if any) should be frozen, how many eggs or embryos (if any) may be donated. Finally, both partners will be asked to read and sign an informed-consent form that indicates they understand the egg-retrieval procedure and the risks associated with it.

ULTRASOUND: THE BEST CHOICE FOR EGG RETRIEVAL

Needle-aspirated egg retrieval under guidance by ultrasound can be performed in a doctor's office, within a specified procedure center attached to the doctor's office, or in an outpatient surgery center environment. It is far quicker, less traumatic, and appears to be more successful than egg retrieval by laparoscopy.

Ultrasound-guided egg retrieval should be performed under sedation and/or with the use of *paracervical block* for pain relief. A paracervical block is a procedure in which local anesthetic is injected on each side of the cervix to numb the nerves surrounding the uterus and cervix. At the same time, local anesthetic is also injected into the upper part of the vagina surrounding the cervix. Some patients prefer to undergo the egg retrieval with the use of paracervical block alone, while the majority will prefer some form of sedation or light anesthesia. (It is advisable that an anesthesiologist be present at the time of the egg retrieval.)

During ultrasound-guided egg retrieval, which almost always is done transvaginally, the physician will introduce a long, sterile ultrasound probe into the vagina. The probe is the projector that transmits the clearly identifiable image of each ovarian follicle to the ultrasound viewing monitor. The physician will then pass a needle via a sleeve alongside the probe through the top of the woman's vagina into the ovarian follicles.

The physician should be able to accurately direct the needle into the follicles by visualizing its progress on the ultrasound screen. In the past, after the contents of the follicle were aspirated, the follicle was usually reinflated with a physiological solution in an attempt to flush out any egg that might otherwise have adhered to the wall of the follicle. Since this added a great deal of time to the procedure, it also created the necessity for prolonged analgesia and/or anesthesia for the patient. In addition, the injection of fluid through the aspiration needle would often reinject an egg lodged in the bore of the needle back into the follicle, making it far more difficult to aspirate a second time.

We believe that this flushing procedure is both redundant and ill-advised. Most important, it hardly ever improves the chances of harvesting an optimal number of eggs. The rare exception might be cases where the woman only has one or two follicles in her ovaries. Under such circumstances the flushing process might ensure that the few eggs present will all be recovered.

Immediately following the procedure, the woman and her partner are informed as to the number and quality of eggs that have been retrieved. The egg retrieval procedure takes about 20 to 30 minutes, with approximately one hour of postoperative recovery. The woman can usually return to normal activity within a few hours of being discharged. The risk is negligible.

LAPAROSCOPY: THE SECOND CHOICE FOR EGG RETRIEVAL

Until the advent of vaginal ultrasonography in the late 1980s, laparoscopy was the preferred method for egg retrieval. It was erroneously believed at that time that the laparoscope enabled the physician to retrieve more eggs.

Laparoscopic egg retrieval is often still performed in association with gamete intrafallopian tube transfer (GIFT). During the laparoscopic GIFT procedure, the eggs are harvested, combined with processed sperm from the male partner, and injected into the fallopian tube(s). However, during the performance of GIFT today, the eggs are usually retrieved through transvaginal aspiration prior to the laparoscopy. In the unlikely event that no eggs are harvested, the laparoscopy is canceled and an unnecessary surgical procedure avoided.

When the woman is fully anesthetized, the physician will usually first make an incision at or near the navel so that a long, thin needle can be inserted into the abdomen. The needle is attached to a gas supply, with which the surgeon insufflates the abdomen with carbon dioxide gas to push the bowel and other organs aside for improved access to the uterus and ovaries.

After the abdomen has been insufflated, the needle is withdrawn and a laparoscope inserted in its place. This long, thin telescope-like instrument is equipped with a high-intensity light source and a system of lenses that enable the surgeon to actually see abdominal and pelvic structures, including the ovaries and the fallopian tubes. One or two additional puncture sites might then be made above the pubic bone so instruments can be inserted to manipulate the ovaries or hold them steady to facilitate the retrieval of eggs from the follicles.

When all these instruments are in place, the surgeon inserts a double-bore needle into the woman's abdomen. Some programs, including ours, use a separate puncture site for this needle because we believe it allows for greater maneuverability. Others insert the needle through a groove or sleeve located alongside a modified laparoscope known as an *operating laparoscope*. Either method is acceptable.

The surgeon inserts the needle into a mature follicle, sucks the fluid out, and places it in a test tube. After the test tube has been carefully labeled, it is taken to the laboratory, where the flushings are immediately examined to determine if an egg was retrieved. The surgeon then progresses from one follicle to another, repeating the same procedure until all or most of the follicles have been aspirated and until as many eggs as possible have been retrieved.

By the time that all or most of the follicles have been aspirated, some of the follicular fluid and flushings will have pooled in the abdominal cavity. The physician will also collect this residual fluid and send it to the laboratory to maximize the number of eggs to be retrieved.

When GIFT is being performed, a number of the most mature and healthy-looking eggs are combined with sperm (previously treated in the laboratory), and the gametes are loaded into a thin catheter. The catheter is then introduced via laparoscope into the outer third of one or both fallopian tubes, where the gametes are discharged.

The laparoscopy usually takes about an hour. Immediately afterward the surgeon will inform the male partner as to how many eggs were retrieved. When the woman wakes up and is coherent (about 30 minutes after surgery), the man usually will be invited into the recovery room to discuss the results with her. (The woman can expect to spend between two and five hours in the recovery room, after which she is usually sent home.)

As the woman will have been told beforehand by the physician and/or nurse coordinator, the gas used to inflate the abdomen may become trapped above the liver under the diaphragm and can irritate the *phrenic nerve;* this produces pain in the right shoulder, right arm, and neck that might persist for 24 to 36 hours. She may also feel abdominal gas pains, and will experience discomfort from the incisions in the abdomen for 24 to 48 hours and sometimes even longer.

Prior to her release from the recovery room, the woman is given instructions about postoperative care, including methods of relieving any discomfort. She should be advised to use only non-aspirin pain relievers, to restrict her diet to light meals for the next day or two, and to notify the IVF clinic or surgeon immediately if she experiences symptoms such as severe bleeding. She should take her temperature twice daily and notify the clinic if it remains elevated over 100°F for longer than 12 hours. In addition, the couple should probably be asked to refrain from sexual intercourse for the next few days.

COMPARING ULTRASOUND NEEDLE-GUIDED AND LAPAROSCOPIC EGG RETRIEVAL

At one time laparoscopy was viewed as the least *invasive* (requiring incisions to be made) of all gynecological surgeries, including egg retrieval. Today, sophisticated non-invasive ultrasound-guided needle-aspiration is performed through the vagina, thereby transforming IVF into a procedure that can be performed either in the doctor's office or in an equipped surgical facility rather than a hospital. Virtually all IVF programs in the United States now retrieve eggs transvaginally under ultrasound guidance in preference to laparoscopy because the ultrasound procedure is less expensive and less traumatic to the patient.

Advantages of Ultrasound-guided Egg Retrieval

1. Because this procedure can be done under sedation instead of surgically under general anesthesia, it results in little discomfort or subsequent incapacitation to the woman.

2. Ultrasound-guided needle-aspiration can be performed in a specially equipped doctor's office or adjacent procedure rooms.

3. Theoretically, since the expense of an operating room is often not incurred and the procedure can be performed relatively rapidly, it could lower the overall cost of an IVF treatment cycle.

Advantages of Laparoscopic Egg Retrieval

1. Because the laparoscope provides direct vision into the abdominal cavity, laparoscopic egg retrieval gives the physician an opportunity to evaluate the anatomical integrity of the pelvis. This makes it possible to determine whether the woman's infertility is amenable to surgical correction, an assessment that might help the couple avoid unnecessary procedures in the future. At the same time, the out-of-pocket cost for infertility treatment could be reduced because in the United States the laparoscopy would frequently be eligible for insurance reimbursement since it was performed primarily for diagnostic purposes.

2. Laparoscopic egg retrieval also affords the physician an opportunity to perform GIFT, which in some centers is still performed in conjunction with IVF. Ultrasound-guided egg retrieval, on the other hand, currently precludes this option, because it does not readily enable the physician to deposit eggs and sperm directly into the fallopian tubes.

3. During laparoscopic egg retrieval it is possible to surgically correct certain anatomical defects that contribute to the woman's inability to conceive. Accordingly, laparoscopic egg retrieval may offer a therapeutic benefit as well as a means of treatment for infertility.

OBTAINING A SPECIMEN OF SEMEN TO FERTILIZE THE EGGS

The laboratory will need a semen specimen for insemination, which occurs four to six hours after egg retrieval. Although most men are able to produce a masturbation specimen upon demand, some may be unable to produce a specimen under the stress of the situation. If it is thought that this might happen, a backup specimen could be collected well in advance and frozen in liquid nitrogen or stored temporarily in special media.

The advantage of collecting a specimen a few days prior to egg retrieval is that the woman can assist her partner by creating circumstances in which he is more likely to be successful. In cases where obtaining a specimen by masturbation is difficult or inappropriate for religious or other reasons, the man can use a special condom while having intercourse so the specimen can be retrieved from the condom; or *coitus interruptus* can be performed.

In select cases it may be necessary to make a small hole at the end of the condom prior to intercourse and to collect the remaining sperm from the condom. The purpose of doing this is to make it possible for a small amount of sperm to pass from the condom into the vagina, thereby removing religious objections from those sects that mandate that semen must reach the vagina for intercourse to be acceptable.

Although it has been suggested that frozen sperm usually fertilize eggs as well as a fresh specimen would, we have found that a fresh specimen is better, especially in cases of male subfertility. We therefore recommend that even if the man has a frozen semen specimen available, he should attempt to produce a fresh specimen around the time when his partner's eggs are to be fertilized.

If the man has very poor sperm quality, it may be necessary for him to produce several daily specimens so they can be concentrated and frozen in case he cannot produce enough on the day of egg retrieval. In such cases it may also be advisable to enhance the quality of the sperm to improve their fertilizing capacity prior to capacitation in the IVF laboratory (see later in this chapter and Chapter 12).

THE LABORATORY'S ROLE IN IVF

The IVF laboratory acts as a temporary womb that supports the delicate gametes (eggs and sperm) and nurtures the newly formed embryos until they are transferred to the woman's uterus.

Nurturing the Eggs and Embryos

Even though the egg is the largest cell in the body, it is too small to be seen in the follicular flushings without a microscope. However, it is usually embedded in the collection of cells known as the cumulus granulosa and collectively termed the cumulus oocyte complex (COC), or corona radiata, which can be seen by the naked eye (see Chapter 2). Once the COC is identified in the follicular fluid, it is examined under the microscope to verify that it contains an egg.

When the egg has reached optimal maturity the entire cumulus mass is placed in a petri dish in a nourishing liquid called an insemination medium. The *insemination medium* is a liquid environment that bathes and nourishes the eggs and embryos just as in nature the woman's body fluids nurture them in her reproductive tract. It contains serum supplements. Each dish is carefully labeled with the couple's name, number, and perhaps even a colored label to guard against any mix-up.

The insemination medium contains a common household product—baking soda, also known as sodium bicarbonate, which maintains the acid-alkaline balance (pH) of the medium at the same level as that found in the body. Without sodium bicarbonate the pH level would fluctuate because eggs and embryos, like any other living cells, convert oxygen, water, and food into waste products and excrete them into the surrounding environment. Sodium bicarbonate neutralizes these acidic and alkaline wastes so they do not threaten the well-being of the eggs and embryos.

Because sodium bicarbonate cannot perform this vital function without an adequate supply of carbon dioxide, the eggs and embryos are kept inside an incubator whose air supply contains a constant 5% carbon dioxide level. Under these conditions the sodium bicarbonate combines with the carbon dioxide to produce the chemical reaction that maintains the proper pH level in the insemination medium.

The eggs and embryos remain inside the incubator for the entire time they are in the laboratory except for brief periods when they are removed to be inseminated, changed to a new medium, or prepared for transfer to the uterus.

Sperm "Washing" and Capacitation

Freshly ejaculated sperm cannot fertilize an egg without undergoing a process called capacitation. In the laboratory, sperm are washed in a special medium to induce capacitation before the insemination of the egg. As explained in Chapter 2, capacitation involves altering the plasma membrane covering the acrosome on the sperm's head; this releases enzymes that will be needed for penetration and fertilization of the egg. Natural capacitation is performed by the fluids in the woman's reproductive tract, especially by the cervical mucus. In the laboratory, capacitation is accomplished by washing and then incubating the sperm at 37°C (body temperature) for about an hour.

The male partner's semen specimen can be prepared for insemination in a variety of ways. One way is to wash the semen in medium and process it in a centrifuge to separate the sperm from the seminal fluid. The sperm gravitate to the bottom of the container, and the seminal plasma is poured off and discarded. Additional medium is then added to the sperm, which are recentrifuged until they again collect at the bottom of the container. At this point the used medium is discarded. New medium is added, and the washed sperm are placed in the incubator to complete the capacitation process.

Frequently, the sperm are allowed to swim through special columns of fluid that contain a substance called Percoll®. The healthiest sperm can in this way be harvested from the bottom of such columns and used for insemination. When male sperm antibodies are present on the sperm, the laboratory may employ columns containing small beads. These columns selectively remove sperm that have antibodies attached to them. In this way it is possible to harvest antibody-free sperm from the column. This improves the chances of successful fertilization in cases where the male partner has a high concentration of sperm antibodies.

Caffeine-like substances are sometimes added to enhance sperm motility—to "wake them up," in effect. And incubation in follicular fluid obtained from one or more of the woman's follicles at the time of egg retrieval or in a special protein medium called *test yolk buffer* may improve the ability of the sperm to penetrate an egg.

Insemination

During *insemination,* the embryologist adds a drop or two of the medium containing capacitated sperm to the petri dish containing the egg. The egg, now surrounded by about 50,000 swimming sperm (the number contained in just two drops of fluid), is returned to the incubator and left undisturbed until the following morning. Fertilization, the actual entry of the sperm into the egg, normally occurs within the first few hours after insemination. In reputable laboratories, each harvested mature egg has a 60 to 70% chance of fertilization. However, if most or all of them fail to fertilize, one might suspect the existence of a previously undiagnosed fertility problem. This is an example of the dual role—both therapeutic and diagnostic—that IVF fulfills.

The Fertilized Egg

About 16 to 20 hours after insemination, the embryologist transfers each egg to a new *growth medium* in order to promote its development and encourage cell division if fertilization has occurred.

The most sophisticated IVF programs examine the eggs at this time to detect the presence of two nuclear bodies (one from the sperm and the other from the egg) within the egg itself. This *pronuclear* stage confirms that fertilization has actually taken place. It also enables the laboratory personnel to select embryos for freezing (*cryopreservation*) at the most favorable time.

By this time the corona radiata cells have condensed around the egg, obscuring its surface and preventing the embryologist from determining whether the egg has been fertilized. In nature, the cumulus granulosa and corona radiata cells are eroded away as the fertilized egg passes through the fallopian tube on its way to the uterus. In the laboratory, these cells must be skillfully removed, or "peeled," to avoid damaging the delicate embryo. *Peeling* is achieved by sucking the fertilized egg and its attached corona into a small-gauge syringe needle or a fine-bore glass pipette and then flushing the fertilized egg out through the narrow opening, thus separating it from the corona.

If *polyspermia* (fertilization by more than one sperm) is detected, the resulting embryo will be discarded because it does not have the potential to grow into a baby. In some cases, however, it is not possible to diagnose polyspermia until later in the developmental process. Then one or more polyspermic embryos might be inadvertently transferred into the woman's uterus. However, this does not cause a major risk to the patient because the polyspermic embryo, like any abnormal embryo, is highly unlikely to implant (see "Miscarriage in Early Pregnancy" in Chapter 2).

More recently, many IVF labs have begun adding cells derived from the growth of other tissue (from the lining of follicles or the fallopian tubes) to the culture medium in which the zygote is being nurtured. It is believed that this technique adds valuable cell-derived, growth-promoting factors to the medium, and that the cells might even in some way direct the embryo to develop in a more healthy manner. This procedure is referred to as *embryo co-culturing* and is currently gaining widespread application.

The fertilized egg (or zygote, as it is now called) will not begin to divide for several more hours. The process of cell division is called *cleavage*. Once the zygote divides, it is known as an embryo.

About 48 hours after the egg retrieval, the embryo is examined under the microscope. By now the cleaved embryo is a translucent, amber-colored mass of four to eight cells. Laboratory personnel often share in the excitement at this stage of the fertilization process, as one usually staid IVF laboratory director confided:

I still get a thrill in the laboratory—even after ten years—when I peel off the corona radiata and find a healthy embryo.

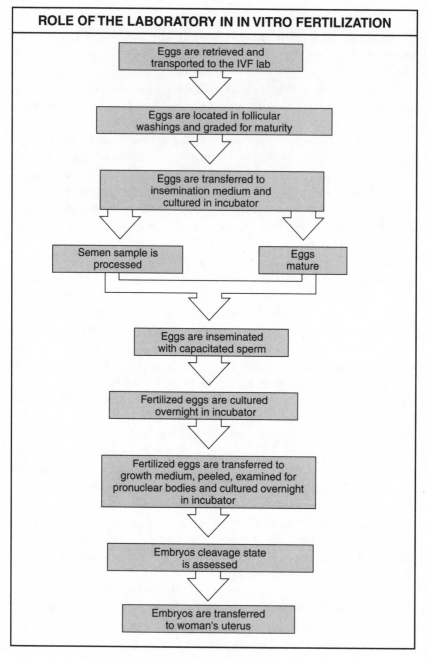

ROLE OF THE LABORATORY IN IN VITRO FERTILIZATION

Eggs are retrieved and transported to the IVF lab

Eggs are located in follicular washings and graded for maturity

Eggs are transferred to insemination medium and cultured in incubator

Semen sample is processed

Eggs mature

Eggs are inseminated with capacitated sperm

Fertilized eggs are cultured overnight in incubator

Fertilized eggs are transferred to growth medium, peeled, examined for pronuclear bodies and cultured overnight in incubator

Embryos cleavage state is assessed

Embryos are transferred to woman's uterus

FIGURE 6-1

Preparing for the Embryo Transfer

Once cleavage has begun, the embryo will continue to divide at regular intervals. (Embryos that divide the fastest are considered the healthiest and the most likely to implant.) Once the lead (fastest-growing) embryo reaches the four- to eight-cell stage, the embryos are usually transferred into the uterus.

Shortly before the transfer, the embryos are put together in a single laboratory dish containing growth medium. The laboratory staff informs the clinic coordinator that the embryos are ready for transfer, and the coordinator prepares the patient and informs the physician that a transfer is imminent.

Figure 6-1 is a flow chart that summarizes the process in the laboratory from egg retrieval to embryo transfer.

CHAPTER

7

IVF STEP 4:
EMBRYO TRANSFER

Although embryo transfer is the shortest step in an IVF procedure, and appears at first glance to be the simplest, it is really the most critical phase of the entire process. Successful clearance of the previous hurdles means nothing if there is a bad transfer with bleeding, pain, and/or damage to the embryos. These problems can occur if too much time elapses between removal of the embryos from the incubator and their transfer into the uterus. A problem-free embryo transfer is so important that we even grade transfers in our setting on the basis of comfort, technical difficulty, and elapsed time so the staff can gauge its success.

Embryo transfer is usually performed between 48 to 72 hours after egg retrieval, but the actual time may depend upon the state of cleavage of the most rapidly dividing embryo or embryos. Most programs prefer to perform embryo transfer before the leading embryo has advanced beyond the eight-cell stage; however, not all the embryos will be at the same stage of division when they are transferred because they develop at different rates.

THE PRETRANSFER CONSULTATION

The couple's first step on the day of embryo transfer is a discussion with the embryologist and the physician, with or without the nurse-coordinator.

The discussion will include the number and quality of the embryos and their cleavage state. The physician probably will reiterate the issues that were raised during previous consultations in regard to the optimal number of embryos to be transferred.

Confronting the Trade-off

The couple will already have discussed with the physician the optimal number of embryos that should be transferred to the uterus. They will know that since the quality of embryos deteriorates rapidly with the woman's age, it is necessary to transfer more—and viable—embryos to offset the quality deficit with quantity. If the woman is under 35 the physician probably has recommended transfer of about four embryos, or six if she is between 35 and 40. In women over 40, larger numbers of embryos may be transferred. Some vacillation over what to do with all the embryos is to be expected up to this point, but now it is time for the couple to come to terms with the success rate vs. multiple pregnancy trade-off as well as the disposition of any remaining embryos.

The physician might say:

Mr. and Mrs. Babcock, all eight of your eggs have fertilized. According to the instructions you gave us before egg retrieval, we are now going to freeze four of them.

At that point the couple have the right to alter their directives as to disposition of their embryos. They may decide to transfer four, or six, or all of the embryos into the uterus. It is their right to change their minds. They may now have decided that the extras should be discarded instead of frozen, or perhaps could be donated to someone else. (It is, however, up to the clinic to decide whether the couple are good candidates for embryo donation.) Or the couple may say that the excess embryos could be made available for experimentation. (Although currently we do not do so, some clinics would use the embryos for experimentation, with the couple's consent.) When the couple have made their final decision, they will complete and sign the consent forms that direct the clinic staff how to proceed.

Reviewing What to Expect During the Transfer

The physician should describe what the couple can expect during the procedure (the details vary widely from clinic to clinic). Some programs, particularly those located in hospitals, perform the embryo transfer in an operating room or a special procedure room, after which the patient is wheeled out on a gurney to a holding area until she has been discharged. Other programs prefer to perform the embryo transfer in the room where the woman will remain for an hour or two.

Some IVF programs allow the male partner to observe the transfer, while others require that he wait in another room. In some settings the man cannot be with the woman until she goes home, even if she has to relax for several hours after the transfer. Other programs, including our own, encourage the couple to be together during and after embryo transfer. We believe this facilitates the bonding process.

The physician should explain that the embryos are transferred by means of a thin catheter threaded through the cervix into the uterus. In programs such as ours, where the depth of the woman's uterus was determined some days prior to the embryo transfer, the woman should be told that the embryos will be transferred a specific depth (just short of the top of the uterus to avoid injuring the endometrium). The intent here is to avoid any injury that might cause bleeding, which has a detrimental effect on the uterus.

Many programs consider that the position assumed by the woman during the transfer contributes to the success of the procedure. Some clinics require that the woman be on her back in the *lithotomy* position during transfer regardless of whether her uterus is tipped backward or forward. Many other programs ask that the woman assume the *knee-chest* position when the uterus is tipped forward so the force of gravity can contribute to optimal placement of the embryos.

THE EMBRYO TRANSFER

Immediately prior to the embryo transfer, the woman should be asked to empty her bladder. This facilitates the transfer procedure and also reduces the likelihood that she will have to urinate while she remains immobilized. She should previously have been asked to avoid coffee or tea because they are natural diuretics and may cause frequent urination. However, she may have been encouraged to eat a light breakfast in order to help combat anxiety. She also should have been asked not to take any medications, including over-the-counter drugs, without clearance from the clinic staff.

It is important that the woman be as relaxed as possible during the embryo transfer because many of the hormones that are released during times of stress, such as adrenalin, can cause the uterus to contract. Some programs believe that imagery helps the woman relax and feel positive about the experience, thereby reducing the stress level. In such a program a counselor and/or nurse-coordinator may help the woman focus on visual imagery for a few minutes immediately prior to embryo transfer so as to enhance her relaxation. As one nurse-coordinator explained, couples may select a variety of creative images to visualize during embryo transfer:

I encourage them to visualize the uterus and the embryo growing within it. Some people imagine little embryos with suction cups on their feet—one woman imagined embryos with Velcro-covered feet. Some people like to visualize a white light coating the baby or a peaceful blue halo surrounding the embryo, or a baby blanket giving the baby a hug, or little soldiers marching into the uterus and digging foxholes in the endometrium. No matter what our counselor or I suggest, the best visualization is what the woman thinks up for herself.

And I'm continually amazed at some of the lucky charms people bring along to the transfer. I have seen them bring drawings and paintings of babies. Some patients wear fertility charms—frogs, African face-mask pendants. Some listen to positive-reinforcement tapes. Women have told me they had worn their lucky dress—and I must confess that sometimes I've worn my lucky dress, too. You can't help getting caught up in this process.

If in spite of appropriate relaxation exercises the woman remains apprehensive, it is often advisable to administer a tranquilizer such as Valium approximately 15 to 20 minutes prior to the embryo transfer procedure in order to enhance relaxation.

When the woman is sufficiently relaxed, she is helped into the appropriate position and made as comfortable as possible. (In programs that rely on relaxation therapy, a counselor or nurse is usually present at the patient's bedside, coaching her in relaxation exercises during the procedure.)

When the woman is in the proper position, the physician first inserts a speculum into the vagina to expose the cervix and then may clean the cervix with a solution to remove any mucus or other secretions. The physician then informs the embryology laboratory that embryo transfer is imminent and awaits the arrival of the transfer catheter that will be loaded with the embryos.

The physician gently guides the catheter through the woman's cervix into the uterine cavity. When the catheter is in place, the embryologist carefully injects the embryos into the uterus, and the physician slowly withdraws the catheter.

The catheter is immediately returned to the laboratory, where it is examined under the microscope to make sure that all the embryos have been deposited. Any residual embryos would be reincubated, and the transfer process would usually be repeated to deliver the remaining embryos.

An embryo-transfer procedure usually takes only a few minutes from start to finish, although it naturally takes longer if the embryos must be reincubated prior to a second or third transfer attempt.

After the Transfer

If the woman is not going to be released immediately after the transfer, she should know whether she can move around freely or if she will have to rest for a specified period. For example, if the woman is in the lithotomy position during transfer, she continues to lie on her back for one or two hours. On the other hand, if she is in the knee-chest position for the transfer, she is urged to remain lying on her abdomen for one or two hours. The male partner or another companion is expected to remain with the patient for emotional support and otherwise tending to her needs.

What the woman should expect to experience physically after embryo transfer is another important issue the physician would have discussed with the couple so they don't become concerned that the transfer has failed. For example, it is not unusual to experience minor cramping or a slight discharge after the transfer. This discharge may merely be due to the emission of fluid retained in the vagina as a consequence of cleaning the cervix in preparation for embryo transfer.

In certain circumstances, distortion of the canal of the cervix due to scarring or the presence of small fibroid tumors or adhesions precludes introduction of the catheter containing the embryos through the cervix into the uterine cavity. The physician will usually have detected this complication during the performance of a mock embryo transfer, which is usually performed a few days prior to the embryo transfer procedure itself (see Chapter 4).

In such a case the physician might attach an instrument to the cervix and gently pull the cervix downwards. This will almost always straighten the canal and readily permit introduction of the embryo transfer catheter. When this is not possible, the embryos are returned to the laboratory and the physician performs a paracervical block to numb the nerves surrounding the uterus and minimize the risk of uterine cramps or contractions during the procedure. Should such cramps occur, they might contribute to the expulsion of the embryos from the uterus. In certain circumstances, additional medications will be given to relax the uterus completely prior to the embryo transfer. Thereupon, the embryo transfer procedure may once again be attempted.

In rare situations where a previously performed mock embryo transfer revealed that it would be impossible to pass the catheter through the cervix, the physician might then elect to perform a *transmyometrial transfer* of the embryos to the uterus. In this procedure, with the patient under general anesthesia, the physician, using a vaginal ultrasound probe, introduces a relatively wide-bore needle transvaginally through the wall of the uterus

into the uterine cavity. The catheter containing the embryos is passed through this needle into the uterus, the embryos are delivered into the uterine cavity, and the needle and catheter are withdrawn.

An Embryo's Chance of Survival

The physician should reiterate that in contrast to a 20 to 25% chance in nature of an individual embryo implanting after it reaches the uterus via the fallopian tube, the probability of implantation following embryo transfer is only 10–15% per embryo.

It was originally thought that this 10% implantation rate occurred because the embryos were lost at the time of transfer and/or the transfer procedure prejudiced the chances of embryo survival and implantation. Recent evidence, however, has shown that healthy embryos derived from the eggs donated by young women have the same chance of implantation as in nature when transferred into a hormonally prepared uterus.

This observation has changed our view on the issues that affect implantation after IVF. It would now appear that some of the hormones produced by the ovary stimulated through superovulation could have an adversarial effect on implantation. Were this not the case, then one would anticipate the same chance of implantation in conventional IVF as with IVF third-party parenting. (IVF third-party parenting will be discussed in Chapter 13.)

In addition, evidence strongly suggests that many embryos do not survive the transfer process either because they have been inadvertently damaged or because they literally get lost. For example, some embryos actually slip back into the vagina when the catheter is removed. They either get lost and die in the cervical canal or vagina, or slide right back out with the catheter. That is why we return the catheter to the laboratory for microscopic examination to ensure that no embryos are adhering to it. Such an examination is well worth the few extra minutes it requires, for in our setting and others as well, reluctant embryos retrieved from the exterior of a catheter have successfully been retransferred into the uterus.

All of these factors working against the implantation of embryos following IVF contribute to the need to transfer enough embryos into the uterus to increase the odds of a viable pregnancy occurring.

The evolution of sophisticated technologies for the performance of embryo transfer might reduce the number of wasted embryos and increase the relatively low implantation rate per embryo associated with currer

embryo-transfer procedures. For example, a group of IVF physicians in Austria have devised a novel approach to securing the embryos' implantation into the uterus and thereby reducing the risk that they will be discharged from the body. These physicians recommend the injection of a natural gluelike substance (derived from a blood protein) along with the embryos to prevent embryo reflux. However, the utility of this technology is doubtful.

THE EXIT INTERVIEW

The exit procedure varies from program to program, but every couple is entitled to an exit interview prior to leaving the IVF clinic. An exit interview prepares and reassures patients for their return home and also provides valuable feedback to the IVF program.

During the exit interview the couple and the physician and/or nurse-coordinator discuss follow-up care, including permissible daily activity, work, travel, and when the couple can resume sexual intercourse. The woman would be advised at this time whether the program recommends hormone supplementation until the pregnancy test confirms or rules out successful IVF. Some programs telephone patients after they have returned home to inquire about their emotional stability and physical well-being. The staff of such programs would likely emphasize that whether or not the procedure has been successful, they would still like to maintain contact with the couple and would be available for consultation at all times.

FOLLOW-UP AFTER THE EMBRYO TRANSFER

The Quantitative Beta hCG Blood Pregnancy Test

Nine days after the embryo transfer, the woman should have a quantitative Beta hCG blood pregnancy test, which can diagnose pregnancy even before she has missed a period. It does so by determining the presence of the hormone hCG, which is produced in minute amounts by the implanting embryo. If hCG is detected, the test is usually repeated two days later in order to see if there has been an appreciable rise in hCG since the first test. A rise (about a doubling of the initial value) usually suggests that an

embryo is implanting and is a good indication of a possible pregnancy. The laboratory then notifies the IVF clinic of the test results.

We usually recommend that these blood samples be drawn ten or twelve days after the embryo transfer. We recommend that both samples be drawn on sequential days, whereupon they will be tested simultaneously in the laboratory. The reason for processing both pregnancy tests on the same day is to enhance the reliability of the results by keeping as many variables constant as possible.

Because an IVF program usually cooperates closely with the referring physician, it is customary for the clinic staff to call the referring physician with the results of the pregnancy test and ask the physician to notify the couple. Thereupon, the clinic staff may also contact the couple. Obviously, if the couple had selected the clinic by themselves rather than on referral, the staff would call them directly with the test results.

Sometimes the physician or nurse-coordinator will work through the referring physician to arrange for the pregnancy tests, and the program may also forward a detailed report about the entire procedure to the referring physician. If the couple wish to make their own arrangements, the program should give them detailed instructions about the necessary tests.

Hormonal Support of a Possible Pregnancy

If the two blood pregnancy tests indicate that one or more embryos are implanting, some programs advocate daily injections of progesterone or the use of vaginal hormone suppositories for several weeks to support the implanting embryo(s). Others, including our own, give hCG injections three times a week for several weeks until the pregnancy can be defined by ultrasound. Some IVF programs do not prescribe any hormones at all after the transfer.

Patients undergoing third-party parenting through IVF surrogacy or ovum donation will receive estradiol and progesterone injections, often in conjunction with vaginal hormone suppositories, for eight to 10 weeks following the diagnosis of implantation by blood pregnancy testing.

We believe that it could, in special circumstances, be beneficial to administer hCG prior to performing the Beta hCG test for pregnancy diagnosis in order to provide better support for the possible pregnancy. This has been practiced selectively in our setting as well as in other IVF programs, and the results are encouraging. The problem with this approach is that administration of hCG prior to the test delays the ability to diagnose pregnancy because hCG is the very hormone that is measured to see if the woman is pregnant.

Confirming a Pregnancy by Ultrasound

Although a positive Beta hCG blood pregnancy test indicates the possibility of a conception, pregnancy cannot be confirmed until it can be defined by ultrasound (see Chapter 10 for a discussion of various definitions of pregnancy). Ultrasound can confirm the existence of a pregnancy, can confirm that it is viable through detection of a heartbeat, and can confirm that it is not ectopic (outside the uterus) four weeks after embryo transfer. The chance of miscarriage progressively decreases from this point onwards. Once a viable pregnancy has gone beyond the third month, the possibility of a pregnancy loss drops to less than 5%.

What to Do if Spotting Occurs

The woman may experience a minimal degree of vaginal bleeding (spotting) after IVF, whether or not she is pregnant. If spotting occurs, she should call her physician immediately and rest in bed. Spotting can be caused by a variety of factors: one of the embryos could be burrowing into the endometrium; one could be attaching while another is detaching; the menstrual period may have begun; or the woman may have a tubal pregnancy. Although there is no way of knowing in the very early stages exactly what is causing the spotting, certain tests can help isolate the problem.

Some women who spot after a positive pregnancy test are undergoing a spontaneous reduction of a multiple pregnancy. For example, many IVF as well as natural pregnancies start off as twins or even triplets and then spontaneously reduce themselves to a singleton or twin pregnancy. Painless bleeding might occur in the process, which could be falsely construed as an indication that the pregnancy is about to be lost.

All women should take the Beta hCG pregnancy test after IVF no matter how much they bleed. One nurse-practitioner explains why:

We had one patient who took her first pregnancy test, but since it was low she didn't take the second one. She kept spotting and spotting, and then went back to her aerobics—three hours every day. After about four weeks I called her up and asked how things were going. She said she had never really had her period and was having these pregnancy symptoms. . . . It turned out that she had conceived in spite of everything!

Women who do not get pregnant but want to make another attempt at IVF should wait until they have had at least one full, unstimulated menstrual cycle in order to prepare themselves emotionally and give their ovaries a rest before the next procedure.

When an IVF pregnancy has been confirmed by symptoms and by ultrasound examination, the woman should seek prenatal care as soon as possible. Thereafter, an IVF pregnancy can be expected to progress no differently from any normally conceived pregnancy given the woman's health, age, and related conditions.

8

WHO ARE MOST LIKELY TO BE SUCCESSFUL IVF CANDIDATES?

Far too many infertile couples gauge the likelihood of their having a baby following IVF on reports from programs that present statistics categorized by the women's ages. While such data certainly provides information by which the relative competency of an IVF program can be assessed, it offers little if any assistance to couples attempting to assess their individual chances of conceiving. We do not believe in assigning arbitrary limits, such as an age cutoff point beyond which we will not accept IVF candidates. We believe it is far more effective to enable each couple to estimate the probability of their conceiving based on how the factors that impact fertility apply to them.

The success rate with IVF depends on a number of factors, many of which have not been clearly defined. Obvious negative factors would include women who do not have a uterus or men whose sperm cannot fertilize an egg. Less apparent are the following critical factors known to strongly influence the infertile couple's chance of successful IVF:

1. The quality of the woman's eggs
2. The number of healthy embryos transferred to the uterus
3. The quality of the uterine lining

There are several other variables including age and the expertise of the IVF laboratory. In this chapter we will describe these factors and variables and explain how they may impact a couple's fertility. In Chapter 9 we will describe the tests that might be administered to both partners to determine the scope of their infertility. These two chapters introduce Chapter 10, which suggests how a couple can use this knowledge to help determine what they can reasonably expect from IVF.

The couple themselves should make their own decisions for treatment based on the reasonable expectations of success they anticipate from a particular approach, given their situation. The doctor shouldn't be expected to do so. And they should be discouraged from accepting the physician's recommendation in place of their own choice.

We believe that physicians should be responsible for giving people information, but cannot make decisions for them. They must explain available options, the medical aspects of any decision that is made, and how couples will be affected by it. Further, resources should be provided so that couples can research the emotional, psychological, and financial issues involved. In the final analysis, it is best they make an informed choice.

VARIABLES THAT AFFECT IMPLANTATION

Age—The Universal Variable

Age is one of the most powerful variables that impact the three ultimatums as well as the other factors. In fact, all forms of infertility treatment, including IVF, are associated with a reduced pregnancy rate in women over 40.

The possibility that IVF may represent the only opportunity of achieving pregnancy for women over 40 stems from the fact that many diseases, such as endometriosis, uterine fibroids, and even tubal disease that can lead to infertility, are far more common in this age category. Moreover, the cost of treatment is likely to be greater for older women because they inevitably require more IVF cycles before there is likely to be a successful outcome.

Infertile couples in which the woman is 40 or over must often overcome two hurdles when they seek fertility treatment: desperation brought about by the realization that time is running out, and discrimination by IVF programs that turn away women over a certain age. Advancing age associated with a progressive decline in a woman's natural fertility understandably induces an overwhelming sense of urgency to achieve a healthy pregnancy before time runs out. Less reasonable is the practice by some programs of turning away older women because their pregnancy rates lower the program's overall statistics so dramatically.

It certainly is true that national statistics show a low IVF birthrate for women over 40. However, we don't believe older couples should automatically assume that success is beyond their reach simply because of the woman's age. To this end, in 1990 we introduced a rigid hormonal and clinical screening process of all IVF candidates to better enable a couple to estimate their likelihood of success. From this assessment, the couple could also decide whether to proceed to IVF. This is far preferable to making decisions based solely on fertility rates reported by age limits or some equally inflexible parameter.

Prior to our using the screening process for women over age 40, the *birthrate per egg retrieval procedure* performed in our San Francisco program in this age category was approximately 10% when IVF was performed in the absence of male infertility. Since implementation of the screening process, our *birthrate per egg retrieval procedure* in the same setting doubled. These statistics justify a nondiscriminatory individualized approach to the performance of IVF in older women.

When such an evaluation indicates a poor outlook for success, most couples will elect to withdraw from treatment or will opt for alternative approaches such as the use of donor eggs or the services of a surrogate. While this is often a painful decision, it does enable couples to get on with their lives and put their futile attempts at conception behind them.

Quality of the Man's Sperm

In the past, sperm quality was particularly important because the poorer the quality of sperm the less likely the couple were to have enough embryos available. However, the introduction of assisted fertilization (AF) through Intracytoplasmic Sperm Injection (ICSI) promises to all but eliminate the negative effect of poor sperm quality on fertilization and pregnancy rates (see Chapter 14). Interestingly, once an egg is fertilized by a man with male-factor infertility problems, the embryo is usually healthy. There is no convincing evidence to suggest that it will be less healthy because a male infertility factor is involved. Thus, three embryos fertilized from a man with a low sperm count should have the same chance of developing into a pregnancy as would three embryos fertilized from a man with normal semen. This variable, of course, is independent of the woman's age.

In our experience, when the man has a concentration of healthy motile sperm of less than 10 million per ml, fertilization ability following conventional IVF (non-ICSI) begins to decline because the sperm's potential for fertilizing seems to be linked to the concentration of motile sperm. The normal concentration of motile sperm in any healthy male is about 50% or greater. Thus, if a man has a sperm count of 100 million per ml, he could expect to have 50 million motile sperm. New laboratory techniques for improving the motility of sperm, along with ICSI, offer renewed hope for men with even the severest forms of sperm dysfunction. In addition, new processing techniques in the laboratory facilitate the removal of some sperm antibodies that also may impact fertilization.

The Presence of Antibodies

Antibodies are proteins that are made by the immune system as a primary defense against infection and injury by foreign substances. The immune system is able to distinguish between that which is "self" and that which is foreign ("non-self"). Antibodies produced in response to foreign proteins and cells will bind to and inactivate the interlopers.

Normally, the immune systems of most women will not react against sperm, but in some cases (for unknown reasons) antibodies are produced that deactivate sperm.

Sperm antibodies. Strange as it may sound, a man can have antibodies to his own sperm. This is quite common after vasectomy, especially if there has been more than a 10-year interval since the procedure was done. The man might have a perfectly normal sperm count, motility, and morphology but the sperm may be coated with antibodies and may even clump or stick together, perhaps even precluding egg fertilization. Special laboratory techniques can partially overcome these problems.

"The immunologic riddle of pregnancy." The immunologic imprint of the implanting *conceptus* (the collective term for the embryo, and the developing fetus and its placenta) is comprised of immunological factors contributed by both the woman and her partner. Since the man's imprint is immunologically foreign to the woman, it is surprising that women don't summarily reject all pregnancies. This is sometimes referred to as "the immunologic riddle of pregnancy." The woman's ability to successfully host a pregnancy depends entirely upon a complex interplay of sophisticated immunologic adjustments designed to convert her uterus to "a privileged site" that would protect the developing conceptus from rejection. But sometimes these mechanisms can go wrong, and depending upon when this happens, the woman might experience repeated pregnancy loss (see later).

In order to better understand the immunologic problems that can compromise or cause pregnancy to fail, we will explain why and how a conceptus is normally tolerated by the mother.

The relevant imprints of the man's immunologic makeup, known as *HLA antigens*, differ substantially from those of the woman. Her immune system recognizes the immunologic difference as soon as the embryo attempts to implant itself into the endometrium. In response, she produces *anti-leucocyte antibodies* (ALA), also known as "blocking antibodies," against the HLA antigens to protect the conceptus from those components of her immune system that would otherwise attack. This quarantines the embryo, protecting it from rejection.

The production of such blocking antibodies is referred to as an *allo-immune response*. In cases where the man and woman share too many of the same HLA antigens, the required allo-immune response is inadequate or fails to take place. Thus, sufficient blocking antibodies do not develop, thereby exposing the conceptus to rejection.

The resulting rejection process may cause the woman's immune system to produce antibodies to certain cell components known as *phospholipids*. This leads to the production of *antiphospholipid antibodies*, which in turn can damage the developing root system of the placenta.

Depending on the severity and timing of the failed allo-immune process and subsequent development of antiphospholipid antibodies, implantation could fail so early that the woman would not even know she was pregnant. In such an event, it is likely that she would be incorrectly labeled as infertile when in reality she was experiencing a failed implantation. Implantation failure occurs far more commonly later on, producing miscarriage, a later pregnancy loss, or retarded growth of the baby (a complication of pregnancy relating to poor development of the placenta).

Auto-immunity. In some circumstances, antiphospholipid antibodies form without the woman having first experienced a pregnancy and/or a failed allo-immune response. In fact, for reasons that remain an enigma, there are conditions where the body simply begins to reject its own tissues out of hand. Examples include rheumatoid arthritis, lupus erythematosis and Hashimoto's disease (auto-immune hypothyroidism). It is not surprising that these are a few examples of conditions that are also commonly associated with repeated pregnancy failure and loss.

The production of antibodies to components of the body's own cells is known as *auto-immunity*. Hence, the conditions referred to above are all auto-immune diseases. The predominant auto-immune antibodies produced in such conditions are also antiphospholipid antibodies.

We now recognize that APA may form in a diverse number of chronic pathologic conditions that damage the cells in the body. This includes conditions such as endometriosis and chronic pelvic inflammatory disease, which are among the most prevalent female causes of infertility that might mandate the performance of assisted reproductive procedures. Clearly, failure to address the auto-immune response that occurs in association with infertility is likely to significantly reduce the chance of a successful outcome regardless of the chosen method of fertility treatment.

We also now recognize that no matter why these antiphospholipid antibodies were formed, they may cause the mother's blood that supplies the early developing placental root system to clot. They also may cause clotting in the blood vessels that supply the placenta, both from the mother's and the fetus's side. This impedes implantation and placental function.

We were recently among the first to recognize and address the important link between auto-immunity and infertility. We in fact discovered that about 50% of infertile women with underlying chronic pelvic infection and 66% of women with endometriosis who presented at Pacific Fertility Medical Centers for ART treatment had antiphospholipid and/or anti-thyroid antibodies. This was in comparison to an incidence of approximately 15% APA in a control group of women of a similar age.

Not surprisingly, we also observed a much lower IVF birthrate in women who tested positive for APA. However, we found that by administering a combination of very low-dose aspirin and a low dosage of the anti-blood-clotting drug called heparin to such APA-positive women, we could more than double the pregnancy rate. Accordingly, we currently test all women for APA prior to their undergoing ART in our settings. If they test APA-positive, we then prescribe aspirin and heparin therapy commencing at the initiation of treatment with fertility drugs. Should the woman conceive, the treatment continues through part or even most of the pregnancy, depending upon the nature and severity of the problem.

Other treatments for immunologic problems relating to reproductive failure. Clinicians and researchers have demonstrated that it is possible in many cases of immunologic infertility and repeated pregnancy loss due to failed allo-immunity to cause the woman to produce ALA/blocking antibodies by immunizing her with leucocytes derived from her partner's blood or from that of a selected donor. This immunization process is repeated at four- to six-week intervals and must be continued throughout most of pregnancy. If successful in eliciting the formation of ALA/blocking antibodies, it would often prevent the rejection process that would lead to the production of APAs.

However, most infertile women who have antiphospholipid antibodies do not acquire these antibodies as the consequence of a failed allo-immune response that results in rejection of the conceptus. Rather, they are there because of an underlying chronic pathologic condition, be it pelvic disease or other systemic causes such as lupus erythematosis and the other diseases in which the body attacks its own cells. Immunization with lymphocytes has no beneficial effect in such cases. It is for this reason that aspirin and heparin therapy is, in the authors' opinion, the mainstay of treating the underlying auto-immune problem that may inhibit many infertile women from achieving pregnancies.

What about the use of intravenous gamma globulin? There are certain circumstances where heparin and aspirin therapy is not sufficient to inhibit the tissue-destructive effect of the auto-immune process. In such cases we advocate the administration of intravenous gamma globulin (IVIG) alone or in combination with heparin and aspirin to combat the damaging effect on the conceptus.

While IVIG therapy at first glance appears to be very appealing, it is quite expensive. In addition, the infusion of blood products is not devoid of the risk of causing local and systemic adverse reactions, and although the possibility of transmitting viral or bacterial infections is very remote, it must always be considered and discussed with the patient. Furthermore, in most cases it would be necessary to administer IVIG every four to six weeks throughout pregnancy. This further compounds the risks as well as the financial burden.

Most important of all is the fact that the majority of women who have infertility in association with auto-immune problems in our experience are responsive to aspirin-heparin therapy. The dosage of heparin administered is not sufficient to cause a defect in the blood-clotting system, and its molecular weight as well as its positive ionic charge precludes its crossing to the baby. Accordingly, it is not associated with birth defects. While aspirin does indeed cross to the baby, more than 100 years of use as well as most long-term studies on the administration of aspirin during pregnancy have failed to reveal any conclusive evidence of a link between the administration of this drug and birth defects or other serious complications of pregnancy.

Other Important Variables

The following additional variables may also impact the couple's ability to conceive after IVF. This list includes physiologically predisposing factors that make a woman more likely to succeed and physical factors such as the technical expertise of a particular IVF laboratory. How they influence the three ultimatums and one another will be explained later in the chapter. These variables are:

  1. The way the woman is stimulated with fertility drugs; the regimen selected for her individually

  2. Genetic defects the couple may carry

  3. Use and abuse of recreational narcotics, such as cocaine and LSD

  4. Pelvic disease due to chronic inflammation, previous surgery, or severe pelvic endometriosis, which sometimes compromise the blood supply to the ovaries and may reduce the amount of functional ovarian tissue. This could influence ovarian response to hormones that induce ovulation and, accordingly, the potential to produce an optimal number of eggs.

  5. Medical expertise of the program. Medical expertise influences the program's ability to evaluate ovarian response and thus the ideal time for egg retrieval. If this process is performed too early or too late, it will adversely influence the likelihood of successful fertilization and hence the number of embryos available for transfer.

  6. Technical expertise of the IVF laboratory. It goes without saying that IVF will not be successful in the absence of optimal expertise in the laboratory. After all, the laboratory is where the eggs are fertilized.

THE QUALITY OF THE WOMAN'S EGGS: THE FIRST ULTIMATUM

Many women over 40 are under the delusion that as long as they are menstruating they can easily get pregnant. It is true they can get pregnant, but it is more difficult for them than for younger women. This is due to the impact of age on egg quality. Age as it relates to egg quality and the other two ultimatums plays a major role in determining whether the woman and her partner will be good IVF candidates.

As a woman ages, so do her eggs. It is not a woman's chronological age per se that so strongly determines the couple's chances at IVF but rather her proximity to the menopause. The closer she is the more difficult it is for her to achieve optimal stimulation with fertility drugs. This is because her lifetime budget of eggs is being systematically reduced by ovulation and/or deterioration as she gets older. And past a certain point, her ovaries are no longer able to produce eggs.

For example, a woman who is going to experience an early menopause around age 40 would react like an older woman in terms of her potential success with IVF. Conversely, a woman who is not destined to go through menopause until age 55 might at 40 respond to fertility drugs the way a 30-year-old would. Since most women experience the menopause around age 50, those over 40 tend to have a significantly reduced IVF pregnancy rate. In fact, we have observed that women under 40 are almost twice as likely to become pregnant with IVF as are those over 40. Thus, chronological age impacts IVF success rates only insofar as it affects the woman's ability to (1) be stimulated with fertility drugs, (2) develop a sufficient number of healthy follicles with eggs, and (3) produce eggs that are fertilizable.

In order to determine the proper dosage of fertility drugs for an older woman, her blood FSH (follicle-stimulating hormone) and estradiol levels should be measured on the second or third day of a natural menstrual period preceding IVF. Studies have shown that the most rapid decline in ovarian function occurs after age 40 and corresponds with a steady rise in the levels of FSH. This is because as ovarian reserve declines, the pituitary gland attempts to reawaken ovarian function by increasing the production of FSH. If FSH is measured during the first, second, or third day of a natural menstrual cycle, its concentration in the blood provides insight into the woman's relative proximity to menopause and helps in determining the ideal dosage and regimen of fertility drugs necessary for her to achieve optimum stimulation.

In other words, the lower the FSH the more likely it is that her ovaries will respond adequately to fertility drugs; and the further she is away from menopause, the lower the dosage and regimen of drugs she would need. Conversely, the higher the FSH the more likely she will need a higher dosage of fertility drugs to stimulate her resistant ovaries.

Approximately four out of every five eggs produced by women under 35 have the potential of producing a healthy baby. A 40-year-old woman will produce healthy eggs approximately half of the time. In contrast, at 45 most of the woman's eggs will be defective. This is probably due to the effect of age on chromosomal integrity. This decline occurs gradually beyond 35, and after 40 it is precipitous. Short of using donor eggs (obtained from a younger woman), there is no known method by which to reduce the increasing incidence of nonviable embryos associated with advancing age.

The higher incidence of birth defects, such as Down's syndrome, in the offspring of older women has been cited in support of the theory of age-related chromosomal damage. It has been determined that 20 to 25% of embryos conceived by younger women naturally or through IVF are chromosomally defective, while the incidence is 50 to 80% in women over 40.

It follows that in order to optimize IVF success rates in older women it is necessary to transfer a proportionately higher number of embryos (usually more than six) to the uterus in the hope of offsetting quality problems. This is because nature mostly rejects unhealthy embryos in favor of implanting healthy ones.

If a woman at 34 has two embryos transferred, at 45 she would theoretically require about 10 to give her the same chance of a pregnancy. Clearly, the fear of multiple pregnancies is something to be considered. But what is important is not the absolute number of embryos transferred into the uterus but the number of *viable* embryos. Again, nature will usually implant only the best ones. Consequently, putting 10 embryos in a woman of 45 would *not* potentially give a greater multiple pregnancy rate than putting two into the uterus of a 35-year-old.

In countries such as England and France, where the number of embryos that a woman has put in her uterus is restricted by law to only three, women over 40 are virtually disenfranchised from conceiving. It is our policy to transfer larger numbers of embryos in women over 40 so as to maximize the chance of implantation. Not surprisingly, given the adverse effect of age on embryo quality, we have not observed an increase in the multiple-birth rate through this approach. In fact, the incidence is three times lower in women over 40 years than for younger women and we have only encountered two sets of IVF triplets in women over 40.

Fertility drugs will only help a woman produce a greater number of eggs, not improve their quality. Thus, egg quality is a limiting parameter that must be met if the couple are to conceive.

THE QUALITY OF THE COUPLE'S EMBRYOS:
THE SECOND ULTIMATUM

The quality of the couple's embryos is directly related to the quality of the woman's eggs. Unfortunately, egg quality can't be upgraded. Therefore, whenever possible we try to compensate for poor quality by transferring as many viable embryos as we can to the uterus, as we outlined in the previous section.

In addition, it is important to understand that women in the oldest age group suffer a higher number of miscarriages (around 40%) once they have achieved pregnancy, because the fetus fails to develop properly due to the poor quality of the egg and, therefore, the embryo.

THE QUALITY OF THE ENDOMETRIAL LINING: THE THIRD ULTIMATUM

The importance of the quality of the endometrium and its potential to promote implantation cannot be overstated. As outlined below, age and the presence of pelvic disease are two factors that are often interrelated. Accordingly, age impacts on the ability of the uterus to support a pregnancy following IVF (and probably also following natural conception).

The endometrial lining of the uterus can be likened to the soil in a planter box. Let's consider the soil in three planter boxes: one with a very thin layer of soil, the second with just an adequate amount, and the third with a thick layer. If the same kind of seed is planted in all of the boxes it is likely that the seed in the first box would be stunted, if it could germinate in the first place. The second might also be somewhat stunted, while the third box would contain a normal, healthy plant.

Compare this with the way implantation occurs in the uterus. When an embryo has cleaved into 30 to 40 cells in the uterus it "hatches" (see Chapter 2). In other words, the zona pellucida cracks and out burst the cells, which try to sink their way into the lining of the uterine wall. Whether this embryo is unable to implant, implants but has stunted growth, or grows into a healthy baby depends on the endometrium, just as the plant's outcome depends on the soil in the planter box.

The endometrial lining may be abnormal in some cases. However, in other cases it is not a matter of pathology as discovered by microscopic diagnosis, but simply an endometrium too thin to permit healthy implantation and growth of an early placenta. A woman with this condition might have little trouble getting pregnant, but might miscarry repeatedly, or lose a pregnancy very early, making it difficult to tell she was pregnant in the first place. A biopsy of the endometrium would likely be completely normal in such cases—there is often a fine line between infertility and recurrent pregnancy loss.

Organisms that negatively impact implantation should be ruled out prior to IVF. Except for ureaplasma, the presence of these organisms is indicated by their symptoms. Ureaplasma rarely presents symptoms, although occasionally it may produce a poor postcoital test (see "Condition of the Cervical Mucus," Chapter 9). The point here is that factors other than the thickness of the lining may result in intractable infertility, recurrent miscarriages, prematurity, stunted intrauterine growth, and compromised quality of life for the offspring.

Much of the original research regarding the factors that influence the quality of the endometrium and its potential to promote implantation was conducted at Pacific Fertility Medical Center. We studied 330 women between the ages of 29 and 45 who collectively underwent 411 completed cycles of IVF. The purpose of this study was to grade the endometrial pattern by ultrasound. We performed vaginal ultrasound examinations prior to ovulation in natural cycles on women scheduled to undergo IVF in a subsequent menstrual cycle. These examinations were repeated a few days prior to egg retrieval in subsequent IVF cycles.

We reached the following conclusion: ultrasound patterns can be measured before egg retrieval to determine whether or not embryos will be able to implant themselves into the endometrium. If certain patterns of the uterine lining are not fulfilled, the woman has little or no chance of achieving pregnancy.

We concluded that a good endometrial pattern determined by ultrasound would be thicker than 8 mm and should appear darker in the middle with a whiter perimeter (the so-called halo effect). Women who do not meet these parameters are about five times less likely to get pregnant. And even if they do, they are far more likely to miscarry.

The good news is that much of the time something can be done to improve poor endometrial lining. Fertility drugs will do so 40 to 50% of the time. If that doesn't help, then special types of estrogen can be administered to further improve the lining. If the woman has poor endometrial lining during a normal menstrual cycle, she would have a 50–50 chance of being able to improve it with fertility drugs. Conversely, if the woman has a good lining in a natural cycle she will almost always respond to fertility drugs by producing a good lining during an IVF cycle.

Among the variables that specifically affect the quality of the endometrium is the condition of the uterine cavity. DES daughters with demonstrable uterine abnormalities, for example, have an endometrium that is less responsive to estrogen, and therefore they will be less likely to conceive. In women who have been on clomiphene (which is derived from DES) repeatedly in numerous consecutive cycles, the endometrium will thin out and production of cervical mucus will decrease. However, this problem disappears after the woman has been off clomiphene for a month.

Fibroid tumors that protrude into the cavity of the uterus might irritate the endometrium and prevent a thick, lush lining from developing in response to fertility drugs or natural hormones. Scarring following an infection after abortion or childbirth, or from an IUD, might also prevent implantation. (The principle behind an IUD is that it, as a foreign body in the uterus, stimulates production of cells (macrophages) that devour an embryo when it reaches the uterus.) Scarring in the uterus could at least theoretically create such an immune-cell response. We believe that surface lesions protruding into the uterine cavity must be excluded before embarking on an IVF cycle. The performance of a hysterosalpingogram is often insufficient to exclude the presence of smaller surface lesions in the uterine cavity. We recommend that a hysteroscopy, under local anesthesia, be performed on all women in advance of their undergoing a cycle of IVF (see "Integrity of the Reproductive Tract," Chapter 9).

Adenomyosis, a condition in which the endometrial glands grow into the uterine wall, creating a spongelike effect, is also associated with poor uterine linings. This condition is sometimes associated with heavy, painful periods and uterine enlargement.

The administration of hormones such as estradiol valerate before starting stimulation is being tried in the hope of priming a better response of the endometrium to the fertility drugs given later.

If the assessment of the endometrium is poor, the woman should probably opt out of that particular cycle of IVF. She then has two choices: (1) her eggs can be removed, fertilized, and frozen for the purpose of transferring them to the uterus in a subsequent cycle when she has been treated with hormonal replacement; or (2) her eggs can be transferred into the uterus of a surrogate (see Chapter 13).

OTHER INDICATORS OF SUCCESSFUL IVF CANDIDATES

The following are additional indicators of potential success in IVF.

The Infertility Is Caused by Female Pelvic Disease

The chances of pregnancy with IVF are highest in cases where the infertility is exclusively due to blocked or irreparably damaged fallopian tubes, and all other parameters are normal.

The Woman Has a Large, Healthy Uterus

The size and shape of the uterus are critical. This is why we evaluate and measure the woman's uterus prior to IVF. As we noted earlier, women who were exposed to drugs such as DES during their own gestation are also less successful with IVF. Finally, women with normally shaped but exceptionally small uteruses have, in our experience, much lower pregnancy rates.

The Woman Has Both Ovaries

Clearly, the ability of the ovaries to produce follicles in response to stimulation by fertility drugs is a priority. The woman's chances of success would be diminished if one ovary or even part of one has been removed surgically, or if she was born with only one ovary. Other complicating factors include ovarian failure (ovaries that do not function properly), diseases elsewhere in the body that may affect the ovaries, and the woman's general hormonal balance.

The Woman Has Had One or More Pregnancies

If the woman has been pregnant before, it can be assumed that her eggs probably will fertilize again. If she has not been pregnant before, inherent genetic or structural egg defects that might not be microscopically detectable can only be identified when fertilization fails to occur in the laboratory.

If she was previously pregnant by her current partner, it is probable that his sperm will fertilize her eggs, but if she has not been impregnated by him it is not possible to determine prior to IVF how her eggs will respond to his sperm.

MEET THE IDEAL IVF CANDIDATES

To sum up, the ideal candidates to undergo IVF using their own eggs would be a couple who answer the following description: a woman under 40, not nearing menopause, who is healthy, ovulates regularly, and has normal hormonal function; whose infertility is due to blocked or irreparably damaged fallopian tubes; who has a healthy uterus of normal size and shape, with a normal endometrial lining, and two normal ovaries; and who has shown that her eggs can be fertilized by having conceived some time in the past—and a man who is perfectly fertile.

The most likely candidates for IVF are couples whose infertility is caused by tubal damage and/or blockage, all other factors being normal. Couples whose infertility is related to male subfertility problems, even if the woman's tubes are normal, have a lower chance of getting pregnant following IVF than do couples in whom the cause of infertility relates to female organic disease.

This does not necessarily mean that couples who do not meet all the non-ultimatum criteria should despair. Depending on the couple's particular set of circumstances, the physician, assisted by a first-rate IVF laboratory, can compensate for the lack of or deficiency in some of the criteria by employing today's high-tech procedures. For example, although both the number and quality of eggs tend to decrease with age, a woman over 40 with a large, healthy uterus and the proper hormonal environment may become pregnant even if the laboratory is able to fertilize only one or two eggs. Chapter 10, "Shaping Reasonable Expectations about IVF," discusses some of the ways in which couples can estimate their IVF outcome although they do not fulfill all the criteria to the letter.

WHEN IS A WOMAN TOO OLD FOR THESE PROCEDURES?

The question of whether a woman is too old for IVF and related procedures raises many social and ethical issues. It is an extremely difficult question to answer, and again, not the kind of question that should be left up to the doctor. It is important that the patient understand that there are risks in pregnancy for women over the age of 40. These risks are compounded by increasing age. A thorough medical assessment can go a long way in helping a woman assess whether she should consider pregnancy. However, there are still some intangibles that will not show up until the woman is pregnant.

The impact of high-tech IVF and other assisted reproduction technology has greatly altered the way many couples—and much of society in general—look at things. Some of the exceptions of 25 years ago are becoming commonly accepted. Times change, and so do attitudes—slowly.

Years ago, if a woman was pregnant over the age of 35, she was often placed in a separate high-risk category. And it was almost unthinkable that a woman over the age of 40 would have a baby. In fact, it was strongly discouraged. Now many women in their forties and even in their fifties are having babies (often through the use of donated eggs—see chapter 13). The point to be emphasized is that in order to get good success rates with IVF it is important to eliminate the variables that might impact adversely on outcome before actually undertaking the IVF process itself.

CHAPTER

9

IS IVF THE MOST APPROPRIATE OPTION?

No couple should attempt IVF until they are certain that no other method of treating infertility would be more appropriate for their particular situation. Because the cause of a problem determines its solution, the couple must first identify the cause of their infertility in order to determine whether they are good candidates for IVF.

Accordingly, a physician with expertise in infertility should examine and evaluate them. The physician will perform certain baseline tests on both partners, and within a month or two should be able to diagnose the problem and prescribe treatment. (As explained in Chapter 3, infertility may be caused by either the woman or the man, or both partners may contribute to some degree.) If the cause cannot be pinpointed, the physician might recommend IVF as a means of both diagnosing and treating the infertility.

WHEN IS THE MAN READY FOR IVF?

The man's fertility must be thoroughly evaluated by all possible methods before the couple resort to IVF. This requires cooperation with the urologist, the reproductive endocrinologist or gynecologist, the laboratory, and the couple. This interactive process has three components: (1) the urologist seeks specific causes for poor semen quality so therapy can be directed at improving it, (2) the laboratory performs tests that evaluate sperm function and often can enhance the sperms' ability to fertilize an egg, and (3) IVF or other assisted reproductive techniques can be used if necessary to help bring sperm and eggs together successfully. Many of the tests that the man can expect to undergo are discussed in the following section.

The Urological Evaluation

The urologist looks for anatomical abnormalities, evidence of obstruction in the scrotum, and physical signs of infection. One anatomical abnormality would be a varicocele, an enlargement of veins in the scrotum, which results in elevated temperatures around the testes that may harm sperm production. In carefully selected cases, tying off or occluding these veins under x-ray visualization may improve semen quality.

When the sperm count is zero or extremely low and the man's FSH level is high, a small biopsy of the testicles may be undertaken to diagnose whether the man is capable of producing sperm. If the sperm count is low and the FSH is normal, the cause may be due to obstruction of the vas deferens or epididymis. If so, microsurgery may be appropriate. And when the urologist finds evidence of infection, antibiotic therapy may be prescribed.

Assessment of the Man's Sperm Function

The reproductive laboratory serves as a link between the urologist and the fertility specialist. The laboratory personnel help establish the diagnosis of male infertility by means of various tests and can treat sperm to enhance their fertilizing capabilities.

Evaluation of male fertility revolves around assessing the quality of the sperm. The sperm's morphology (percentage of normally shaped sperm), configuration, motility (ability to travel through the reproductive tract and fertilize an egg), and count (number produced in a semen specimen) can be determined under the laboratory microscope.

The word *morphology* is derived from the Greek term for shape or structure. Sperm that appear to be normally shaped are more likely to fertilize an egg, while those with obvious structural abnormalities are less likely or perhaps cannot do so at all. Fortunately, it is possible for the man to be fertile even when the percentage of normal sperm is low.

If the sperm count or motility initially appear to be abnormal, the physician should determine whether hormonal problems are the cause. This might require extensive blood testing to evaluate the function of the thyroid gland, the output of the pituitary gland, and the secretion of sex hormones by the testicles into the man's blood. While no single medication can improve sperm counts, abnormal hormone levels may justify the use of fertility drugs to improve the semen. Fertility hormones such as clomiphene citrate, hMG, and hCG have been used with varying success in such cases.

Sophisticated testing may be necessary if the problem cannot be readily identified. For example, if the morphology, motility, and count appear normal but the sperm do not swim properly, or if some sperm huddle in clumps, immunologic studies on the man's blood and sperm might be necessary to determine whether he is producing antibodies that are inhibiting sperm function, as mentioned in Chapter 8. This sometimes happens if infection, injury, or surgery have damaged the sperm ducts or testicles.

The Zona-free Hamster-Egg Penetration Test

A technique known as the *zona-free hamster-egg penetration test* helps determine whether the sperm are likely to fertilize healthy eggs. In this test, the sperm are placed in a laboratory dish along with hamster eggs from which the eggshell-like zona pellucida has been removed. Then the number of sperm that penetrate the eggs are counted and evaluated. Failure to penetrate a sufficient percentage of eggs may indicate severe male infertility. Although hamster eggs can be penetrated by human sperm, they will not cleave and develop into viable embryos.

One of the drawbacks of the zona-free hamster-egg penetration test is that it may produce misleading results. For example, a small percentage of women whose partners failed the hamster test subsequently get pregnant. In other cases, the test results will appear to be normal although the man's fertility is severely impaired. Because this test is highly technical, with a high potential for error, various laboratories may interpret it differently, with the conflicting results just described.

Unexplained infertility is caused in many cases because some physical-chemical, biochemical, or immunological factors within the egg's surrounding cumulus complex prevent the sperm from undergoing the acrosome reaction (being attracted to and then penetrating the zona pellucida). The only way to determine whether these inhibiting factors exist is by observing the interaction of human eggs and sperm in the petri dish during IVF.

Because the zona-free hamster-egg penetration test is a somewhat imperfect screening test of the ability of a man's sperm to fertilize his partner's egg(s), a negative report should not necessarily deter a couple from attempting IVF.

The Hemi-Zona Test

The *hemi-zona test* uses eggs from surgically removed ovaries or from human cadavers. It is possible to render such eggs unfertilizable by bisecting them or by aspirating their contents prior to performance of the test. The eggs are then exposed to sperm. By observing the sperm interaction with the eggs, laboratory personnel are able to determine whether the sperm are capable of attaching to or penetrating the surface of human eggs and, thus, whether the acrosome reaction can take place. The hemi-zona test augments but does not replace the zona-free hamster-egg penetration test as a screening method for male-factor infertility. Moreover, the availability of cadaver eggs limits the feasibility of this test.

Treatment of Male Infertility

Male infertility can be treated by surgically repairing anatomical defects in the reproductive tract or by administering hormones to stimulate the testicles. In rare situations, a disease outside the reproductive tract that hinders sperm production may be corrected medically or surgically. For example, if the man is producing antibodies to his own sperm, steroids may be used to suppress this immunologic reaction (but steroid treatment itself is relatively ineffective and is not without risk).

Only if the cause of the man's infertility is not correctable through minor surgery or administration of hormones should a couple consider IVF. We state categorically that because of dismal success rates in cases of severe male infertility, intrauterine insemination is inadvisable in such cases.

The performance of gamete intrafallopian transfer (GIFT) is, in our opinion, also contraindicated in cases of male infertility because while it might provide a reasonable chance of pregnancy occurring, it does not afford any diagnostic ability. In other words, because in GIFT the eggs and the sperm are placed in the fallopian tube prior to fertilization, unless the woman conceives following the GIFT cycle there is no way to assess the fertilization potential of her eggs. In vitro fertilization offers both the optimal diagnostic and therapeutic opportunity in such cases.

As we noted in the previous chapter, once an egg is fertilized, the resulting embryo probably has the same chance of implanting as would one from a couple whose cause of infertility is not a male-factor problem. This chance is roughly 10 to 15% per embryo transferred. We want to reemphasize here that the probability of successful fertilization occurring either within the petri dish or within the fallopian tube is lower when male-factor infertility is present. The extent to which fertilization is affected will depend upon the severity of the sperm defects. As the total number of normal, motile sperm decreases, the fertilization rate also declines. However, the introduction of ICSI (see chapter 16) could change all that. Now it is possible to achieve the same fertilization and birthrates in cases of severe male infertility as with indications for IVF. Furthermore, even men who produce *no sperm* at all in their ejaculates can have a small biopsy taken from one of their testicles or have sperm aspirated from a blocked sperm duct, followed by ICSI on their partner's eggs, and father their own children.

WHEN IS THE WOMAN READY FOR IVF?

As in the man's case, defining the cause of a woman's infertility requires careful analysis of her history and a clinical examination by a fertility specialist. Three areas need to be carefully assessed: (1) the pattern of ovulation, (2) the anatomical integrity of the reproductive tract, and (3) the status of the cervical mucus. We will discuss these issues below along with many of the tests that can be used to evaluate and sometimes correct deficiencies in these three areas.

The Pattern of Ovulation

As explained in Chapter 2, a woman is unlikely to conceive unless she ovulates at the right time in the proper hormonal environment. Reliable evidence of ovulation can only be obtained by examining the woman's ovaries through a laparoscope, by repeated ultrasound examinations around the time of presumed ovulation (to detect a collapsed follicle and/or some follicular fluid pooled in the abdominal cavity) or—the only really reliable way—by confirming a pregnancy with ultrasound.

It is important to remember that the purpose of assessing ovulation is not simply to determine whether the woman ovulated but whether she ovulated at the proper time, within the proper hormonal setting, and if the lining of the uterus is properly prepared. While the following tests will not provide conclusive evidence that the woman has ovulated, they will help the physician determine how well synchronized these vital components are during the luteal phase (second half) of the menstrual cycle. These tests are valuable indicators of the appropriateness of ovulation and the hormonal environment when their results are considered together.

The BBT chart. One way to assess the pattern of ovulation is by compiling a daily body temperature chart. The woman does this by taking her oral temperature each morning before she gets out of bed and before she has anything to drink. She then charts her temperature on a *basal body temperature* (BBT) *chart.*

The pattern on the chart will mirror the output of the hormone progesterone by the corpus luteum in the ovary. A woman's temperature begins rising about 12 to 24 hours following ovulation, and this rise will be sustained through the rest of the menstrual cycle (when the ovaries are producing hormones to prepare the endometrium). In other words, the woman's temperature goes up when progesterone is produced and stays up until its levels drop with the demise of the corpus luteum. This phenomenon occurs because progesterone acts on the biological thermostat in the brain that regulates body temperature, setting it one notch higher when progesterone is being produced and lowering it at or just before the beginning of the menstrual period. As long as the temperature is up, it can be assumed that progesterone is present.

Because ovulation usually occurs 14 days before the expected menstrual period, the woman's temperature will usually be about ½° to 1° higher during the last two weeks of her cycle. A biphasic pattern such as this (the first phase is lower than the second phase) suggests ovulation. It is not important whether the increase is sudden or gradual.

Figure 8-1 shows how the biphasic BBT pattern is synchronized with hormone production, ovulation, and other aspects of the menstrual cycle. Estrogen levels in the blood (including estradiol, as shown in Figure 8-1) peak at the time of ovulation. This triggers the LH surge and to a lesser extent a synchronous rapid rise and fall in the FSH levels. The woman ovulates 8 to 36 hours after LH is first detected in the urine by home ovulation testing. About a day later the temperature rises. The estrogen level falls progressively immediately prior to and for a few hours following ovulation and then rises again, only to drop precipitously immediately prior to menstruation. Measurable amounts of progesterone first appear in the bloodstream around the time of ovulation and then escalate during the second half of the menstrual cycle and, as with estrogen levels, also drop sharply preceding the onset of menstruation.

All of these hormonal interrelationships are orchestrated by the follicle, which contains the developing egg and produces estrogen. The follicle begins to grow during the first half of the menstrual cycle. Then immediately after ovulation it collapses and forms the corpus luteum. It is the corpus luteum that during the second half of the menstrual cycle produces both estrogen and progesterone. The corpus luteum has a natural life span of about 12 to 14 days, whereupon its failure causes the abrupt fall in estrogen and progesterone levels cited above. This results in a withdrawal of the hormonal support of the endometrium, thereby precipitating menstruation. Should pregnancy occur, the corpus luteum's survival is prolonged, estrogen and progesterone levels continue to rise, and menstruation is deferred (see "Hormones Prepare the Body for Conception," Chapter 2).

The BBT chart, therefore, (1) indicates that the hormonal events associated with ovulation have occurred and that the woman presumably released an egg from her ovary, and (2) provides a rough idea about when ovulation occurred and the length of the second half of the cycle. Nothing more should be read into the chart, however. It is a misconception that the BBT chart can precisely pinpoint the ideal time for intercourse.

The cervical mucus begins to thicken rapidly a few hours following ovulation and becomes almost impenetrable to sperm within 24 to 48 hours after ovulation. As the temperature rises around the same time, the likelihood that pregnancy will occur when natural intercourse is timed with the temperature rise is remote. Therefore, the old adage about the woman saying "Hurry home, honey, my temperature has risen and we should try and have a baby!" is without foundation. By that time, it will already be too late to get pregnant. It is far better to time intercourse with the onset of the LH surge as determined by urine testing, as the detection of LH in the urine precedes ovulation by 8 to 36 hours, a time when the condition of the cervical mucus should be optimal.

Regular menstrual periods. Regular menstrual periods associated with breast tenderness and, often, mood changes preceding the onset of the periods plus some discomfort (cramping) during menstruation are all clinical indications that the woman is likely to be ovulating.

Urine tests. Since the surge of the hormone LH triggers ovulation, detection of LH in the urine suggests that ovulation has probably occurred. The woman can easily take this urine test at home around the time of ovulation by performing one of the commercially available "home ovulation tests" on a sample of her urine. The actual time of ovulation can be predicted by daily, twice daily, or even more frequent performance of the test and charting of the results. It should be kept in mind that there is a two- to four-hour lag before the urine test will reflect the surge of LH in the bloodstream (see Figure 8-1).

The LH blood-hormone test. An LH blood-hormone test done several times daily around the time of presumed ovulation (at least 14 days prior to the menstrual period, or when the temperature goes up) is a relatively sophisticated indicator of likely ovulation. This is related to the fact that the LH surge in the middle of the cycle precedes ovulation.

When women used to undergo IVF in natural, unstimulated cycles or when they were given clomiphene citrate without hCG (which acts like an

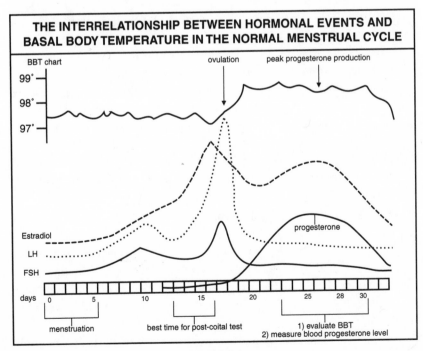

FIGURE 8-1

artificial surge of LH), it was necessary to measure LH by repeated blood tests. This series of tests was necessary in order to pinpoint the time of expected ovulation so the eggs could be retrieved before they were ovulated into the fallopian tube or abdominal cavity. Today, the dipstick urine tests that measure the excretion of LH (described in the previous section) are accurate enough. Serial LH blood testing, which is expensive, time-consuming, and painful, has been virtually supplanted by serial urine LH testing to pinpoint ovulation.

Endometrial biopsy. An *endometrial biopsy,* which can be performed in a doctor's office, evaluates the condition of the lining (endometrium) of the uterus. In this relatively painless procedure the physician inserts a small curette or suction apparatus into the uterus a few days prior to expected menstruation and removes a sliver of the endometrium for examination under the microscope.

If the endometrium is secretory, indicating the effect of the progesterone, it is likely that the woman has already ovulated. A pathologist is frequently able to pinpoint almost to the day when ovulation is likely to have occurred by microscopically examining the secretory endometrium.

It is also possible to determine whether the endometrial sample is appropriately developed for the stage of the menstrual cycle at which it was taken, thus providing yet another parameter of whether ovulation is occurring in the proper hormonal environment. It is important to remember that although estrogen is not measured in this test, progesterone can act only on a lining that has been stimulated by estrogen. So if microscopic examination shows that the endometrium development is in sync in the middle of the luteal phase, it can be assumed that estrogen stimulation was adequate as well. By combining the endometrial biopsy with measurement of blood or urine hormone levels, it is possible to further refine the evaluation of ovulation.

The progesterone blood-hormone test. The most common indicator of presumed ovulation is the level of the blood hormone progesterone in the woman's blood, as measured a few days before her anticipated menstrual period. This test usually will indicate whether or not the woman is likely to have ovulated, again because progesterone is typically only present in significant amounts in the bloodstream after ovulation.

Analysis of both the LH and progesterone values measures not only whether these hormones are present but also the appropriateness of their levels in the blood relative to the time of the cycle (see Figure 8-1).

The bottom line. It is important to view all these tests in context. For example: (1) the blood progesterone level of a week or so prior to menstruation may suggest that ovulation has occurred; (2) a microscopic examination of an endometrial biopsy done a few days prior to menstruation might show that the endometrium has responded appropriately to progesterone; and (3) the length of the luteal phase (as gauged from the urine dipstick test or temperature rise on the BBT chart to the onset of menstruation) was 13 to 14 days. This would indicate a good luteal phase (proper length, enough progesterone, and histologically a normal uterine lining).

The opposite situation might be, on day 25 of the cycle: (1) an endometrium typical of day 19, (2) a significantly lower-than-normal progesterone level a week before menstruation and (3) a luteal phase of about nine days. This would suggest possible corpus luteum insufficiency and/or insufficient stimulation of the endometrium. In this case the luteal phase is too short, the endometrial response is inappropriate for the timing of the cycle, or the hormonal environment is inadequate. The presence of any of these parameters will lead to the diagnosis of an abnormal hormonal environment. Administration of fertility drugs such as clomiphene, hMG, and hCG will often regulate the pattern of ovulation so that most women with such problems can have a reasonable chance of conceiving. Therefore, ovulation problems alone do not require IVF as a solution.

Integrity of the Reproductive Tract

Since pregnancy can occur only when the reproductive system does not inhibit the passage of sperm, eggs, and embryos, an abnormality within the tract is likely to interfere with the woman's ability to conceive. Careful analysis of her medical history, including any previous venereal diseases, infections following the use of an IUD, and the results of previous fertility tests may suggest the presence of an abnormality such as damaged fallopian tubes, endometriosis, or fibroid tumors. However, defects in the reproductive system can only be confirmed through tests such as those outlined below.

Hysterosalpingogram (HSG). The *hysterosalpingogram* tests the patency (absence of blockage) of the fallopian tubes and the shape of the uterus. During the HSG procedure the physician injects a dye into the uterus via the vagina and cervix. The dye shows up as a shadow on an X ray. The HSG can identify a blockage in the fallopian tubes and may also point out some obvious abnormalities inside the uterus. These could include fibroid tumors, scarring, and abnormalities of uterine development that might occur spontaneously or as a result of DES exposure during the woman's own prenatal development.

The major shortcoming of the HSG is that its usefulness is limited to assessing the interior of the uterus and fallopian tubes. In addition, it does not yield any information about the ovaries. In other words, fertility problems in which the inside of the fallopian tubes and uterus may appear perfectly normal but where the fallopian tubes are unable to retrieve the eggs from the ovaries because of surrounding scar tissue or ovarian tumors cannot be diagnosed by HSG. Also, an HSG cannot diagnose endometriosis, in which the fallopian tubes are usually open.

Women who undergo an HSG should be aware that the procedure is relatively painful, often causing severe cramping. In addition, there is about a 2% risk that the dye could convert a dormant infection into a full-blown pelvic inflammation. Women should also be aware that infection can be introduced if the procedure is not performed using the proper sterile technique. Finally, because non–water-soluble dyes tend to collect in damaged tubes and produce infection and complications, there has been a move away from using such dyes. However, the iodine solution in both water-soluble and non–water-soluble dyes can also cause reactions.

Laparoscopy. One way to both assess the patency of the fallopian tubes and examine the pelvic cavity is by inserting a laparoscope through a small incision in the abdomen. Similar to the lighted, telescope-like instrument used during IVF- or GIFT-related laparoscopy, it enables the physician to look inside the abdominal cavity and perform surgery at the same time. The physician can directly visualize the patency of the fallopian tubes during laparoscopy by injecting a colored, water-soluble (non-iodine) solution into the uterus through the vagina and cervix, and observing the fluid's passage through the fallopian tubes. It is also possible to microscopically examine the ends of the fallopian tubes through the laparoscope. Conditions such as endometriosis and pelvic adhesions can be diagnosed with confidence only by laparoscopy or laparotomy, which often afford the opportunity for the physician to treat the problem surgically at the same time.

Diagnostic laparoscopy. *Diagnostic laparoscopy* is a procedure that can be performed under local anesthesia in a doctor's office with minimal discomfort for the patient; or it can be carried out at the time of a therapeutic laparoscopy, under general anesthesia.

Falloposcopy. In a new procedure known as *falloposcopy*, the physician can direct (by watching through the laparoscope) a thin fiber-optic catheter through a separate puncture site into one or both of the fallopian tubes at the time of laparoscopy and then examine the tubal lining for adhesions or damage. Falloposcopy even enables the physician to perform surgery inside the fallopian tubes through the falloposcope, thus avoiding the need to open the abdominal cavity except for the small puncture site. During the same procedure the physician can examine the ovaries as well as the exterior of the fallopian tubes and uterus to see if endometriosis, inflammation, or any other anatomic defect might be contributing to the woman's infertility. However, this procedure is not commonly available.

Because of its ability to both assess tubal patency and enable the physician to visualize the abdominal cavity and treat most cervical conditions at the same time, laparoscopy has largely replaced the hysterosalpingogram as the most popular method of assessing the anatomical integrity of the reproductive tract.

Diagnostic hysteroscopy. *Diagnostic hysteroscopy* is an office procedure that is performed under intravenous sedation, general anesthesia or paracervical block with minimal discomfort to the patient. This procedure involves the insertion of a thin, lighted, telescope-like instrument known as a hysteroscope through the vagina and cervix into the uterus in order to fully examine the uterine cavity. The uterus is first distended with carbon dioxide gas, which is passed through a sleeve adjacent to the hysteroscope.

The diagnostic hysteroscopy facilitates examination of the cervical canal and the inside of the uterus under direct vision for defects that might interfere with implantation. Conditions such as fibroid tumors, polyps, bands of scar tissue, or congenital abnormalities can readily be detected. We have observed that approximately one in eight candidates for IVF has lesions that require attention prior to undergoing IVF in order to optimize the chances of a successful outcome. We strongly recommend that all patients undergo hysteroscopy to correct the pathology prior to IVF.

Therapeutic hysteroscopy requires general anesthesia and should be performed in an outpatient surgical facility or conventional operating room where facilities are available for *laparotomy,* a procedure in which an incision is made in the abdomen to expose the abdominal contents for diagnosis, or for surgery should this be required.

Therapeutic hysteroscopy is usually combined with laparoscopy. This allows confirmation that the uterus was not perforated during the removal of surgical lesions and that surrounding structures have not been traumatized.

Surgical procedures to correct defects can be done hysteroscopically. This usually requires prior distension of the uterus with a fluid. Laser-directed procedures can also be performed through the hysteroscope, thereby reducing the chance of bleeding. Therapeutic hysteroscopy has eliminated much of the need for major abdominal surgery, with its incumbent risks.

Reparative surgery to correct anatomical defects. In most cases, the physician would first attempt to correct defects in the reproductive tract by the least traumatic form of surgery, probably through the laparoscope, salpingoscope, falloposcope, or hysteroscope. Today almost all pathologic conditions of the uterus, fallopian tubes, or surrounding structures that compromise fertility are accessible to laparoscopic or hysteroscopic surgical treatment. With the notable exception of women under 35 who have had a tubal ligation for sterilization purposes, any anatomical defect (causing infertility) unable to be corrected using the laparoscope or hysteroscope probably requires IVF. Laparotomy for tubal disease is fast becoming a thing of the past.

The advisability of corrective surgery on the fallopian tubes depends on the situation. If a young woman has had a tubal ligation for the purpose of sterilization and now wants her fallopian tubes reconnected, tubal microsurgery may be her best option, depending on whether (1) the entire tube was destroyed in the process, (2) there is enough of the tube remaining on either end to allow reconnection, (3) the fimbriated ends of the fallopian tubes are intact, and (4) previous surgery has caused scarring around the tube that inhibits normal egg pickup by the fimbriae.

Condition of the Cervical Mucus

Cervical mucus insufficiency causes about 5% of all infertility problems. Some of the causes of cervical mucus problems include previous surgery for cervical cancer, which sometimes destroys the mucus-producing glands of the cervix, and excessive freezing (cryocautery) of the cervix because of lesions, early cancer, or exposure to drugs such as DES while the woman was inside her mother's womb. Women with DES exposure will likewise also be unable to produce a lush endometrial lining in spite of normal ovulation and normal blood estrogen levels prior to ovulation.

Mucus evaluation at ovulation. The simplest way to evaluate the cervical mucus is to examine the woman around the time of presumed ovulation. Since the ovulation tests discussed previously will help to pinpoint the expected time of ovulation, it is possible to estimate when to conduct this examination. At the appropriate time the physician inserts a speculum in the woman's vagina, retrieves some cervical mucus, and evaluates the physical-chemical properties that normally occur in the mucus at ovulation. Healthy cervical mucus should be clear, stringy (stretchy), and should form a fernlike pattern when dry.

At the same time the physician may conduct a *postcoital test* (PCT), or *Hühner Test,* to assess the interaction of the mucus and sperm. The woman should have been asked to have intercourse from 6 to 18 hours before her appointment. Then the physician will take a mucus sample from her cervical canal via a catheter and examine the mucus on a slide under the microscope.

The PCT assesses the number of sperm in the mucus and evaluates their motility. The presence of a large number of sperm moving in a linear and purposeful fashion usually indicates that the mucus is healthy. The PCT is a good screening test because it provides a rough evaluation of the quality of the sperm as well as of the cervical mucus. At the same time, a less favorable postcoital test does not necessarily indicate a cervical-mucus deficiency because the sperm could be at fault. It should be borne in mind that the most common causes of a poor postcoital test are the following:

1. Poor timing of the test. If the test is performed more than 12 hours after ovulation has occurred, the cervical mucus may have become thick and tenacious, thereby preventing the invasion of sperm. Accordingly, the postcoital test should be performed immediately prior to or at the time of ovulation. The best method of ensuring that the timing is correct is by performing daily home ovulation urine testing, which will show a color change 8 to 36 hours prior to the occurrence of ovulation.

2. Male infertility. When the male partner has a low sperm count there will be a tendency for a postcoital test to have a poor outcome.

3. The presence of infection of the cervical mucus with ureaplasma, chlamydia, or other organisms may also produce poor postcoital test results.

4. The presence of sperm antibodies in either the woman's or the man's secretions will thwart the postcoital test.

5. Destruction or absence of the cervical glands because of surgery or infection will limit the amount of cervical mucus produced and also adversely affect its quality, thereby yielding a poor postcoital test.

6. The use of clomiphene citrate for more than three consecutive months and possibly the use of clomiphene citrate in women over 40 tends to be associated with a poor-quality cervical mucus, and a poor postcoital test will result.

7. A DES abnormality of the cervix is commonly associated with poor mucus production.

Tests of the cervical mucus and blood for antibodies. If analysis of the cervical mucus at ovulation and a postcoital test are not sufficient, the physician may look for sperm antibodies in the cervical mucus and blood. In such cases it may be necessary to culture the mucus or analyze the blood for various microorganisms that might have destroyed the cervical glands' ability to produce the proper mucus.

TUBAL SURGERY AND IVF—APPLES AND ORANGES

In general, the forms of sterilization that can best be reversed microsurgically are those in which a small portion of the fallopian tube was either blocked or cut away at one place in the midportion of the tube. In such cases, there is a 70 to 80% chance that patency of at least one fallopian tube can be reestablished successfully through tubal reconstructive surgery (*tubal reanastomosis*). The subsequent birthrate is approximately 50% within two to three years after the procedure.

Restoration of tubal patency after sterilization gives the woman a permanent chance of conceiving. Therefore, this is one of the few situations in which we might recommend tubal surgery over IVF—provided the woman undergoing the procedure is under 35.

It should also be recognized, however, that a single cycle of IVF performed in the optimal setting could afford the same woman a greater than 30% chance of having a baby from a single attempt provided tubal occlusion was the only reason for her infertility problem. IVF is, of course, far less invasive than a laparotomy or laparoscopy and does not require general anesthesia, hospitalization, or a protracted time off work.

Two or three attempts at fertilization in an optimal IVF setting would almost certainly result in pregnancy rates that greatly surpass those that could be obtained through the performance of tubal reconnection. Moreover, if such a woman desired to retain her contraceptive ability after IVF has produced a baby, her fallopian tubes would still be occluded and her contraception intact. In contrast, following tubal reanastomosis the woman will have to use some form of contraception.

Another point to be considered is that a 3% tubal or ectopic pregnancy rate occurs following IVF pregnancy as compared with a 15 to 20% risk following the performance of tubal reanastomosis.

For the above reasons, it is essential that the opportunities offered by both tubal reanastomosis and IVF be carefully discussed with the couple so they can make an informed decision.

The reason for the relatively high success rates following tubal reanastomosis is that aside from a small segment of the fallopian tube that is surgically occluded at the time of sterilization (and is removed when reanastomosis is performed), the remainder of the tube is almost always normal. In contrast, the fallopian tubes are damaged to a lesser or greater degree through disease in almost all other cases where tubal surgery is performed. This explains the difference in success rates referred to in this section.

In our opinion, tubal surgery will be likely to facilitate pregnancy only in situations where the fimbriated ends of the fallopian tubes are normal or can be restored through surgery to a relatively normal anatomical configuration. Without functional fimbriae, it is highly unlikely that an egg would find its way from the ovary into the fallopian tube. For example, a *salpingostomy*, an operation in which the end of a blocked fallopian tube is surgically reopened (often eliminating the fimbriae), provides relatively little hope for the infertile couple. In such cases, the woman often stands less than a 20% chance of having a baby within two years of tubal surgery, whereas the same woman would have better than a 40% chance after a single attempt at IVF performed in an optimal IVF program. In addition, major surgery carries with it prolonged hospitalization, the risk of complications, increased cost, lost time away from work, incapacitation, and significantly greater discomfort.

A better birthrate of about 40 to 50% can be expected within two to three years of removing adhesions from around the fallopian tubes as long as the fimbriae and the insides of the tubes are otherwise normal. Younger women (under 35 years) may still choose surgery over IVF in such cases, provided they are willing to allow two to three years for the surgery to have a good chance to work. It is, however, a reality that more than 70% of women who undergo tubal surgery for the treatment of infertility will ultimately require IVF.

As long as insurance companies in the United States reimburse for about 80% of the costs of tubal surgery but are unwilling to fund IVF, and as long as consumers remain uninformed about the benefits of IVF, most couples will still select tubal surgery over IVF for financial reasons. Two-thirds of all tubal surgeries performed in the United States are still being done in cases where for all practical purposes the fallopian tubes have been irreparably damaged. In the remaining one-third, tubal surgery is more appropriately performed to free adhesions around the fallopian tubes and ovaries, remove fibroid tumors, and treat endometriosis.

Actually, comparing tubal surgery with IVF is a futile apples-and-oranges exercise. How can a procedure such as IVF, where success rates are determined on the basis of a single menstrual cycle of treatment, be compared with tubal surgery, in which evaluation of the success rate per procedure requires a wait-and-see approach that often spans two or more years? And tubal surgery and IVF still cannot be compared on an equal footing in today's economic climate because insurers are not likely to reimburse for IVF although they almost always will for tubal surgery.

In order to equate the two procedures it would be necessary to balance the success rate achieved through several attempts at IVF over a period of a year or two against the quoted success rate following tubal surgery over the same time span. Three or four attempts at IVF, performed even in the average IVF setting, are likely to result in a higher success rate than any form of tubal surgery.

The insurance environment surrounding IVF in the United States can be expected to change (although gradually) if current legislative and judicial trends continue (see Chapter 15). As more lawmakers and courts direct insurance carriers to fund IVF, the popularity of IVF will inevitably surpass that of most forms of tubal surgery. We predict that when the financial burden is eliminated, most informed women would choose several attempts at IVF performed over a period of a year or two in preference to undergoing major tubal surgery, with its incumbent risks, associated pain and discomfort, and protracted period of convalescence.

Table 9-1 compares risks and benefits of IVF and tubal surgery.

Table 9-1

TUBAL SURGERY VS. IVF

	Tubal Surgery	IVF
Risk of complications	+	+/-
Cost (out-of-pocket)	+/-	++
Time off work	++	+/-
Pain/discomfort	++	-
Chance of successful pregnancy (excluding tubal reanastomosis)	+/-	++
Risk of ectopic pregnancy	++	+/-

+ = moderate
- = almost absent
++ = great

As one successful IVF patient who had previously undergone tubal surgery explains, making the cost-benefit trade-off in favor of IVF was a good decision:

I went through three attempts at tubal surgery to open up my fallopian tubes. The first time I had a salpingostomy, but the tubes closed up again and the doctor did a second operation to try to reopen them. My third surgery was to break down the scar tissue around my fallopian tubes, which at that time were open. This was five years ago. If I had only realized that my chances were poor from the beginning I might have considered IVF earlier. There's far less pain and discomfort with IVF than surgery, and—finally—I'm expecting my first baby in a few months.

This does not mean, however, that all major pelvic reparative surgery should be avoided regardless of the condition of the fimbriae. Severe lesions that involve the uterus, ovaries, or fallopian tubes and threaten the woman's health often require surgical correction. In such cases the emphasis should be on protecting the woman's well-being first, and then addressing her infertility.

IVF SHOULD BE CAREFULLY WEIGHED AGAINST OTHER OPTIONS

We strongly believe that any relatively noninvasive, nonsurgical technology that is applicable to the couple's particular problem is preferable to an attempt at IVF. (See Chapter 14 for a detailed discussion of new technologies that might be undertaken before IVF.)

Factors that should be considered when deciding between alternative procedures and IVF include: (1) success rates of the various procedures; (2) financial considerations; (3) physical toll; (4) emotional investment; and (5) the insistent ticking of the biological clock, which may mandate quick action if ovarian failure for the woman is imminent. With other alternatives, especially surgical options, a woman is usually given a long-range hope: "Now that you've had your operation, let's see how you do in the next two to three years." There is always the possibility that during those 24 to 36 months she will not conceive.

Gradually coming to terms with failure in such circumstances is an easier, slower adjustment than it would be with IVF. With IVF, the time frame is compressed into one month, and when a couple fail to conceive during one treatment cycle they know immediately they have nothing to show for their efforts. The emotional impact is far greater with IVF. It is abrupt and painfully traumatic. Accordingly, couples should decide whether other alternatives that require a lengthy period of anticipation might be more appropriate for them before embarking on IVF.

But when couples choose IVF over other assisted reproductive technologies, the rewards can be great. The following comments from two first-time parents in their mid-forties describe how repeated failures at pregnancy shaped their attitude toward IVF:

Husband: *When we started trying to have a baby about 15 years ago there wasn't anything like IVF available. We went through all the different fertility tests, and sometimes we gave up temporarily and then decided to try again. We had a lot of disappointments.*

Wife: *If it hadn't been for my husband I would never have tried IVF because I was so tired of being let down. But he said if you don't try you'll have nothing. It was hard emotionally to go through IVF, to relive those hopes and risk getting hurt again, but I'm so glad I did. We have a three-month-old daughter now who has dramatically changed our lives.*

10

SHAPING REALISTIC EXPECTATIONS ABOUT IVF

Procreation—and with it the ability to achieve immortality by living on through one's children—is an inalienable right as well as one of the most insatiable human needs. This strong natural urge exerts tremendous pressure on couples unable to have a baby. And the pressure to reproduce becomes increasingly acute as couples grow older and become more aware of their own mortality.

Although IVF offers hope to many infertile couples who until recently had no way of conceiving, it is not a panacea for every couple who want a baby. In addition, every IVF procedure exacts an emotional, physical, and financial price from both partners. No one gets through the process without paying the toll. All couples considering IVF should learn what they can reasonably expect from it before they commit to the procedure. Once they have shaped realistic expectations about their probable experiences, they are ready to decide whether IVF is truly for them.

WHAT ARE A COUPLE'S REALISTIC CHANCES OF SUCCESS?

The following profile fits the ideal candidate/couple for IVF:

1. The woman is under 40.

2. She has at least one healthy ovary capable of responding normally to fertility agents (a low FSH/E_2 on the second or third day of the cycle).

3. The uterus and the endometrial lining are normal and capable of sustaining a healthy pregnancy.

4. The woman has been pregnant in the past, thereby proving that her eggs can fertilize.

5. The man's sperm function is normal.

These criteria are not absolute, however. For example, in many cases the adverse effect of advancing age on egg and hence embryo quality could in part be offset by transferring a greater number of embryos to the woman's uterus, thereby permitting natural selection to determine which embryos are healthy enough to implant. Alternatively, the application of advanced technologies such as assisted fertilization through intracytoplasmic sperm injection (ICSI) micromanipulation of the egg or eggs might promote fertilization in cases associated with sperm dysfunction (see Chapter 14); or assisted hatching (see Chapter 2) might improve implantation potential in cases where the zona pellucida enveloping the embryo is too thick or resistant, as is now believed to occur with advancing age.

Here is a practical example. Let's say a woman is 35 years old, her partner has normal sperm function, and following stimulation with fertility drugs eight eggs are retrieved from her ovaries. She has a normal uterine cavity, as assessed by hysteroscopy, and an excellent endometrial lining, as evaluated by ultrasound. Such a woman would, in an optimal IVF setting, have better than a 40% chance of getting pregnant per IVF cycle of treatment.

Now, if the same woman had a male partner with sperm dysfunction, but by employing ICSI the physician was able to transfer four to six embryos to her uterus, she could be expected to have the same chance of getting pregnant regardless of male infertility. High technology, by improving the chances of fertilization, has offset the adverse effect of sperm dysfunction on outcome.

Here is another example. Let's say the woman is 41 years old with exactly the same optimal circumstances as the younger woman described above. Her chances of pregnancy would be significantly reduced because of the adverse effect of age on egg and embryo quality. Her chances of getting pregnant would be about 20% if four embryos were transferred to her uterus and roughly 30 to 40% if double the number of embryos were transferred. Here, an age-related embryo-quality deficiency has been partly offset by transferring a greater number of embryos to the uterus.

It could be argued that she might have a strong chance of a multiple pregnancy if so many embryos were put into her uterus. But this is not the case, because most of her embryos would be deficient and thus would be rejected by her uterus through the natural selection process. Effectively, only three out of the eight embryos might be healthy enough to produce a baby. (The effect of age on the multiple pregnancy rate has been discussed elsewhere in this book.)

Therefore, couples who are contemplating IVF should initially base their realistic expectations for success on those criteria that have the potential to predict outcome, and then temper their expectations with the realization that many of the adverse factors can be partially overcome. However, it is not always possible to overcome severe deficiencies. For example, embryos may never be able to implant into a uterus with surface lesions such as fibroid polyps that protrude into the uterine cavity or severe scarring in the uterine cavity due to previous infection. Such conditions would interfere with implantation.

Each couple and their physician, then, must assess these criteria, given their own unique set of circumstances and the environment of the IVF program they have selected, to determine their own realistic expectations of getting pregnant.

THE COUPLE MUST BE SURE THEY ARE TRYING TO CONCEIVE FOR THE RIGHT REASONS

Both partners owe it to themselves, each other, and their unborn child to examine what they expect to achieve from parenthood and what they are willing to contribute before they get on the IVF roller coaster. Successful parenting is usually rooted in a stable relationship and a sincere desire for children. An infertile couple who embark on IVF without first considering their motivation can end up with problems even worse than the infertility problem that brought them into IVF in the first place. The addition of a baby to a troubled relationship will compound existing problems as well as create new ones.

Some childless couples get caught up in the pursuit of pregnancy because they have an idealized picture of what it would be like to have a baby. They are so intent on proving they can conceive that they lose sight of how their lives will change when a baby becomes part of the family. They may not fully consider whether they are willing to adapt their lifestyle to accommodate a child's demands on their time, energy, mobility, and financial situation—from babyhood on.

The couple must be sure they are not trying to have a child to please someone else—a mother who has always wanted a grandchild, for example. Each partner should feel comfortable knowing that it is okay not to have children at all if either or both prefer to be child-free. Granted, it is often difficult to come to terms with family and societal pressures, but it can be done.

But when a couple want a baby for the right reasons, family and friends are likely to share in the joy over the baby's birth. As one new IVF mother reported:

We've had nothing but total support from our entire family. They call Caryn the "miracle baby." Friends I went to high school with have called or sent cards, and people I don't even know have sent gifts. We're very proud that we went through IVF and that she's here.

The woman's mother expressed her feelings about her daughter's IVF experience with these words:

My daughter tried for seven years to get pregnant. I was beginning to feel I was never going to have a grandchild, and I worried through the whole nine months of her pregnancy. But IVF was marvelous for all of us. It was very exciting, very satisfying.

The woman's father, who was concerned about the procedure's effect on his daughter, was won over after the baby's birth:

My daughter had undergone tubal reconstruction and laparoscopies and so much pain that I thought if IVF failed it would be so defeating for her. I wondered if it was going to be worth it. Of course now I can see that it was well worth it. I just didn't want to see her undergo any more traumas.

And the new mother's best friend told us:

When you have friends who have had infertility problems for years, you suffer with them. You feel guilty about the pleasure you're having from your own family every time they come to visit. So we all feel like Caryn is our baby. She's theirs, but they still have to share her because we've shared all the pain with them.

Couples who choose IVF for the right reasons are likely to reap the greatest rewards from parenthood. They are often the most committed of all parents because they are making the greatest sacrifice to have a baby.

IVF IS AN EMOTIONAL, PHYSICAL, AND FINANCIAL ROLLER-COASTER RIDE

The biggest decision an infertile couple will ever make in regard to IVF is whether or not they really want to become parents. Once they agree that they are committed to parenthood, they must next decide whether they are ready to deal with the emotional, physical, and financial consequences of their actions.

Both Partners Must Share in the Emotional Cost

An IVF procedure requires an enormous emotional commitment at each level of the program, whether or not IVF is successful. This has a permanent impact on the couple. Because the toll can be so great, both partners must be committed to supporting each other from the very beginning.

Based on the statistics reported by our program when this book was being written, 36% of women under the age of 40 who undergo IVF at our program in San Francisco will likely have a baby following a first attempt at egg retrieval, provided the male partner is fertile. For women between 40 and 43, the comparable success rate is about half that number. The success rate falls so sharply after age 43 that such women would probably be best advised not to undergo IVF with their own eggs.

The chance of a baby being born following conventional IVF using donor eggs at the Pacific Fertility Center in San Francisco (provided the donor is younger than 35, the recipient has a healthy uterus as determined by hysteroscopy, and the male partner is fertile) is about 50%, regardless of the birth mother's age. The success rates for the second or third attempts are about the same across the board. We have observed that success rates begin to decline from the fourth IVF attempt onwards.

In the past the existence of sperm dysfunction roughly halved the anticipated success rates in all age categories and circumstances. However, since the introduction of ICSI, many centers of excellence (including our own) are achieving the same success with IVF performed in cases associated with even the most severe degree of sperm dysfunction as when the procedure is undertaken using normal sperm. ICSI has leveled the playing field. It should be borne in mind that some women who fail on the first three tries do get pregnant after four or more attempts. Therefore, it is realistic to be optimistic—guardedly, cautiously optimistic. But the couple should also be realistic enough to prepare themselves emotionally so they are not overwhelmed by failure in case IVF does not succeed.

The IVF Procedure Is Stressful

Both partners should be prepared to respond to a variety of emotionally stressful demands as they undergo IVF, including

1. Dealing with general stress "baggage" (shame, guilt, anxiety, depression, anger) they bring into the program because of their long-standing battle with infertility

2. Following new procedures; interacting with a strange and sometimes impersonal clinical staff, and perhaps with a constantly changing cast of characters

3. Living in an unfamiliar environment: new state or country (many couples will travel from another state or country to undergo IVF in a good program), different daily schedule, time-zone changes, separation from their normal support network

4. Coping with the unpredictable emotions that the fertility drugs trigger in the woman

5. Reacting to family and marital stress, which may be heightened by the constant need for mutual support

6. Managing the financial aspects of the procedure

Couples react to the demands of IVF in strikingly different ways. One expectant mother thought the stimulation phase of her second IVF treatment cycle (her first cycle had ended in an ectopic pregnancy) was the most stressful:

One of the most difficult things I went through was the roller-coaster ride waiting for the estradiol level. Would it be high enough? Would I have enough eggs? Would I have to be on another day of fertility injections? It was really the most exhausting part of the entire process.

Fortunately, she produced three eggs and had two embryos transferred (as opposed to five during the first attempt), and she was six months pregnant at the time of this interview.

The mother of a one-month-old IVF son also found the waiting to be most trying:

IVF Also Makes Physical Demands on the Couple

The physical demands of IVF range from the annoyance of hormone shots and blood tests to the discomfort of egg retrieval for the woman and the need for the man to produce a semen specimen on demand. The couple probably will have undergone a variety of diagnostic procedures to determine the reason for their infertility and thus may already be familiar with some of these demands.

Certainly when compared with tubal surgery, the process of ultrasound egg retrieval presents minuscule risk, discomfort, and complications. However, it is still an emotionally and physically draining experience. In addition, if the couple have selected a program in another state or country, they may undergo additional physical discomfort as a result of the stress of travel, including jet lag and the general disorientation caused by temporarily living in unfamiliar surroundings.

Proper emotional preparation and mutual support throughout the treatment cycle will help both partners cope more effectively with the physical demands on the woman. And they should keep in mind that once the pregnancy is confirmed, the remainder of the gestational period will probably vary little from pregnancies experienced by all other expectant women.

IVF Requires a Significant Financial Commitment

Until IVF is universally funded by medical insurance, it will continue to be a program for the haves, not the have-nots in the United States. This is true even though its cost only really becomes significant when the woman is wheeled into the operating room for egg retrieval. That is when the fees for anesthesiology, surgery, processing and fertilizing the eggs and sperm, and transferring the embryos mount into thousands of dollars within a few days. Until the egg retrieval, the couple will only have to pay relatively minimal costs.

However, the cost of attempting to conceive is not usually limited to the IVF procedure. Many couples have learned how high the overall expenses of attempting to conceive can be. As one newly expectant IVF patient said:

So far we've spent about $43,000 trying to get pregnant, so the IVF portion was really a minor part of the total cost. I first went through reconstructive surgery, then five or six laparoscopies. I shudder to think of the money we spent on airfare to consult with doctors in other cities, plus hotel rooms and meals, to say nothing of all the income we lost by taking so much time away from work. Had we known that my tubes were permanently blocked, we could have saved a lot of money by going *directly through IVF. But out of that $43,000 our insurance company has paid about $13,000, so we have been pretty lucky financially.*

Although this woman considered herself lucky to have paid "only" $30,000 out of her pocket, a similar outlay would be prohibitive for most other couples. That is why couples contemplating IVF should first determine whether their budget can accommodate all the direct and indirect expenses that IVF entails.

A Realistic Attitude Toward Budgeting for IVF

When inquiring about the costs of a particular IVF program, the couple should always ask for written quotations that cover all the charges they will incur. This is to avoid any distressing surprises that might be caused, for example, if the program omitted a hefty charge for multiple tests from its quotation. The couple should not be afraid to ask about items they do not understand.

In addition to the total fee quoted by the program, the couple who must travel from their locality to a program elsewhere should budget generously for the kinds of expenses mentioned above: air or ground transportation, meals, hotels, other travel expenses, allowances for lost income, even baby-sitters or house sitters while they are on the road. If they plan to visit several IVF programs before selecting one, they should also allow for the expenses they will incur during each site inspection.

In many cases, the couple will have to pay up front for the entire procedure. IVF programs usually request payment in advance in order to avoid problems collecting from the couples who do not get pregnant. Most programs will advise the couple how to bill the insurance company for the reimbursable components of the procedure but will not bill the company directly.

A Realistic Attitude Toward Insurance Coverage

In vitro fertilization candidates should not automatically assume that their insurance will cover IVF. Reimbursement practices vary from company to company and from state to state. As we have mentioned throughout this book, the attitude of insurance companies in regard to IVF could be vastly improved. As one new mother said vehemently:

We're still waiting for our insurance to pay. It's been over a year since we went through the IVF program, and they just keep making excuses. So far we've only received $380!

The father of triplets expressed his concern about the unfairness of insurance companies that refuse to fund IVF but cover other surgical procedures without question:

Through all of our infertility treatments, including artificial insemination and surgeries, the insurance companies argued and refused to pay. Then our triplets were born seven weeks premature, and the hospital bill for them and my wife was $128,000! The insurance company said that was no problem and that they were going to pay the whole thing.

The financial risk in IVF is great, but the return can be priceless. That is why it is so important for each couple to be absolutely sure of their willingness and financial ability to make such an investment before they attempt IVF. Yet more and more couples are willing to make the financial commitment. Why? When asked if he and his wife had difficulty deciding whether to undergo IVF given its cost and uncertain outcome, one new IVF father responded:

Well, when you really want children you set your priorities. We think babies are more important than fancy vacations or a sailboat. We were able to budget for IVF. But we're sorry that insurance doesn't usually cover it because a lot of people just can't spend $10,000 or so to go through these procedures.

In Chapter 15 we go into more detail about why most insurance companies still do not reimburse for IVF.

HOW MANY TIMES SHOULD A COUPLE ATTEMPT IVF?

Because of the emotional, physical, and financial toll exacted by IVF, it is preferable that no one undertake a one-shot attempt. If a couple can only afford one treatment cycle, IVF is probably not the right procedure for them. After all, there is only about one chance in three that IVF will be successful—and a tremendous letdown if it fails.

We believe it is unreasonable to undergo IVF with the attitude that "if it doesn't work the first time, we're giving up." In vitro fertilization is a gamble even in the best of circumstances. But statistically speaking, the couple who have selected a good IVF program are likely to have a better than 60% chance if they undergo IVF three times, as long as their gametes can fertilize, and the woman is under 40 and has a normal uterus and a proper hormonal environment.

Unfortunately, some people are destined to remain childless. In our opinion, it is rarely advisable to undergo IVF more than four times in a reputable IVF program that gets good results. After that, the time has probably come to consider other options, such as ovum donation, IVF surrogacy, or adoption. One woman, who eventually adopted a newborn boy, described her disappointment over three failed procedures:

It's very difficult to deal with. You go into any of these procedures with the expectation they will work. Somehow we are raised in our society to think that it's not whether you are going to have children, but how many do you want? We plan for our car, and we plan for our house—and assume that the children are going to come. And when they don't, it's devastating. You are basically out of control of your own body. There is nothing that you can do to make the egg and sperm unite.

Couples who choose to undergo IVF should realize from the outset that the inability to become pregnant should never be considered a reflection on them as individuals. They should view the entire procedure with guarded optimism but nevertheless must be emotionally prepared to deal with the ever-present possibility of failure.

COUPLES MUST CONSIDER THE POSSIBILITY OF A MULTIPLE PREGNANCY

As we explained in Chapter 4, the couple who are unwilling to settle for a low pregnancy rate must be prepared for the possibility of multiple offspring. With IVF, twins are born in about one out of every four pregnancies and triplets in one out of 30. This compares to twins once every 80 births and triplets once every 6,000 in nature. It is important to distinguish between the number of multiple *births* (being discussed here) and the number of multiple *pregnancies* (see Chapter 4). This is because most women who are carrying more than twins will opt for pregnancy reduction. Also, many multiple pregnancies reduce spontaneously.

Because the incidence of larger, more hazardous multiple pregnancies is higher with IVF, the couple should be familiar with the concept of selective pregnancy reduction. This trade-off between pregnancy rate and the possibility of multiple births is one of the most important realistic expectations that couples must determine.

SOME COUPLES MAY HAVE MORAL/ETHICAL/RELIGIOUS OBJECTIONS TO IVF

While a discussion of the moral, ethical, and religious dilemmas created by IVF is not within the scope of this book, we would encourage all couples to come to terms with their concerns in this regard before entering an IVF program. No one should be excluded from an IVF program because of religion any more than because of age, color, marital status, or sexual preference. Every case should be assessed on its own merit. The couple should be willing to discuss their concerns openly with their physician; the IVF program staff; and their minister, priest, or rabbi. Sometimes, by working together, it is possible to find approaches that will satisfactorily resolve everyone's concerns.

MATCHING REALISTIC EXPECTATIONS WITH THE RIGHT PROGRAM

Once the couple have formed their own realistic expectations about IVF and have decided to undergo the procedure, they are ready to select a program. When evaluating potential programs the couple should expect an IVF provider to meet three basic criteria: (1) the program should provide the highest quality of medical care; (2) the program should ensure that the couple will have the best possible opportunity of conceiving within the guidelines of sound medical practice; and (3) the program should deal with the couple in a manner consistent with the emotional, physical, and financial investment they will make. The following chapter explains how consumers can evaluate IVF programs on the basis of these characteristics.

CHAPTER

11

HOW TO FIND
THE RIGHT IVF
PROGRAM

The infertile couple should begin their search for the right IVF program by talking with their own physician and/or a local fertility support group. If there are no fertility support groups in the area, the couple should contact the national headquarters of one of these organizations. They may also wish to talk to couples who have already undergone IVF, as these couples tend to develop a close network and will probably be happy to share their experiences. Contacts made through such networking will likely lead to even more sources of information.

We wish to stress that no matter how strongly the couple feel that time is closing in on them, it is important to devote a few months to diligent research rather than rushing arbitrarily into the most convenient program.

HOW SHOULD THE SUCCESS OF AN IVF PROGRAM BE EVALUATED?

The process of selecting an IVF program is significantly different from that of choosing a gynecologist, whose credentials alone assure the couple of his or her competence and expertise. First, there is no accrediting agency that provides information to consumers about an IVF program's competence and success.

The couple should evaluate more than the expertise of just one person. They should also take into account the success rate of all the individuals who operate as a team. For example, a laboratory that is not very successful at fertilization would be a drawback in a program that has a friendly, supportive staff and otherwise presents a reliable, innovative image.

How does one gauge the success of an IVF program? In the broadest terms an IVF program's success can be measured by its

1. **results**—a track record that is consistent with currently accepted rates for successful IVF procedures;

2. **caring**—the degree to which the couple perceive an attitude of caring manifested by the staff;

3. **staff interaction**—whether there seems to be open, harmonious interaction among the staff involved in the program, and whether the couple feel comfortable dealing with the staff; and

4. **reputation**—how the program is regarded by those who have undergone IVF, by the community in which it is situated, by other physicians—and don't forget the program's financial stability.

The only basis for judgment the couple will have when trying to select the most appropriate IVF program is their observation of the way the program operates—from initial contact until patient discharge. In order to properly research individual IVF programs, they will first have to learn to understand and interpret the terms and statistics they will probably encounter.

HOW IS PREGNANCY DEFINED?

The word *pregnancy* often means different things to different people. For example, the terms *chemical pregnancy* and *clinical pregnancy* are frequently used interchangeably although they have completely different meanings. It's necessary to understand both of these definitions in order to avoid misinterpreting the statistics that may be quoted.

Chemical Pregnancy

Chemical pregnancy refers to biochemical evidence of a *possible* developing pregnancy. A positive blood or urine pregnancy test confirms a chemical pregnancy provided that the woman has not received the hormone hCG recently and does not have a tumor that releases hCG into her blood (see Chapter 7 for a discussion of the quantitative Beta hCG blood pregnancy test).

Clinical Pregnancy

A clinical pregnancy is one that is *confirmed* rather than merely *presumed*, as with a chemical pregnancy. A pregnancy can be confirmed when evidence of gestation either in the uterus or fallopian tube is detected by ultrasound, and/or when pathological evidence of placental or fetal tissue is obtained following miscarriage or surgery. A blood or urine test alone is not sufficient to confirm a clinical pregnancy.

Chemical vs. Clinical Pregnancy

The couple should keep in mind that only 25 to 30% of all natural pregnancies survive long enough to postpone the menstrual period, thereby creating even a suspicion that the woman is pregnant. This means that most chemical pregnancies never become clinical pregnancies.

Verifying a chemical pregnancy when a woman has undergone IVF is complicated by the fact that she has almost invariably received an injection of hCG 12 to 14 days prior to the pregnancy test. Depending upon her body's absorption and excretion rates, small amounts of the hCG may still be present in her blood at the time of the test and could result in a false suggestion of a pregnancy.

Therefore, the term chemical pregnancy when applied to IVF rates might mean one of three things: (1) a true chemical pregnancy is present but will not progress to a clinical pregnancy (the most likely scenario), (2) a chemical pregnancy is in the process of developing into a clinical pregnancy, or (3) the result was a false indication of a chemical pregnancy caused by residual hCG.

If the terms chemical pregnancy and clinical pregnancy are used interchangeably, a quoted pregnancy rate could be falsely inflated, perhaps by as much as 100%, by citing the percentage of chemical pregnancies for that particular program. For this reason most reputable IVF programs will not report chemical pregnancies in their statistics.

Consumers should be aware that some programs report "inclusive pregnancy rates," which are clinical and chemical pregnancy rates combined. However, since it is not always possible to determine which terms are actually being quoted when this reporting method is used, couples should not be afraid to ask the proper questions to clarify and distinguish between these two terms.

HOW SHOULD A REPORTED PREGNANCY RATE BE INTERPRETED?

Careful study and reference to the following two sections on success rates should enable you to ask the hard questions of the programs you are considering.

Pregnancy Rate per Embryo-Transfer Procedure

This refers to the clinical pregnancy rate per embryo transfer performed. Quoting the pregnancy rate on the basis of embryo transfer will inflate the overall results because if no eggs are retrieved or if eggs are retrieved but do not fertilize in the laboratory, the woman accordingly does not undergo an embryo transfer. Thus her case, which actually represents a failed IVF procedure, will not be reflected in the statistics. Had her case been included in the computations, the overall rate would be somewhat lower.

Pregnancy Rate per Number of Women in the Program

Some IVF programs base their pregnancy statistics on the number of women who undergo the IVF procedure. But to report the pregnancy rate in this manner will further inflate results because this method fails to allow for women who may undergo more than one IVF procedure. If the number of patients rather than the number of egg-retrieval procedures performed is used as the statistical base, the success rate will naturally look better.

Pregnancy Rate per Attempted Egg Retrieval

We believe that the pregnancy rate quoted to consumers should be the number of clinical pregnancies that occur per egg retrieval. Once the woman undergoes the egg retrieval, she is having the procedure, and that should be the basis for arriving at this statistic.

The variation in statistics that these three definitions can produce is startling. Take a program that is experiencing a 20% rate of clinical pregnancies based on the number of egg retrievals performed. If the program instead reports clinical pregnancies on the basis of embryo transfers, its reported results could be 25%. Now, if it bases its statistics on the number of clinical pregnancies per number of women patients, it might (depending on the number of times each patient has undergone IVF) quote a success rate of as high as 30%. No wonder it is so important to be able to interpret these statistics intelligently. Imagine how these rates could be manipulated even further if the program were to include chemical pregnancies in its computations.

The term *cumulative pregnancy rate* is often used to describe the overall chance of a clinical pregnancy occurring per egg retrieval or per embryo transfer following several successive procedures. *Cumulative birthrate* refers to the overall chance of a woman having one or more babies per egg retrieval or per embryo transfer following several attempts.

WHAT IS AN ACCEPTABLE SUCCESS RATE?

Birthrate per Egg Retrieval Procedure

While proper reporting of a program's pregnancy rate per egg retrieval reflects the competence of a program, the only statistic that really matters in the final analysis is a woman's chance to have one or more healthy babies per IVF procedure.

We believe that an acceptable success rate is one that is at least as high as the average of the rates being reported at that time from all programs that have experienced a clinical birthrate. All of the IVF programs certified by the Society of Assisted Reproductive Technology (SART) have experienced a clinical birthrate. Because the mean birthrate per egg retrieval in this country is about 20% for women under 40, based on the statistics released by SART, it is fair to say that a program offering a 20% or better chance of a baby being born is operating within the realm of current acceptability.

If the mean birthrate per egg retrieval of 20% should increase to 25% in the next year, then 25% should become the new acceptable minimum level. The birthrate per egg retrieval will not be 20% forever because many programs are improving. However, consumers can expect that the average success rate will increase very slowly.

Anticipated Birthrate per Egg Retrieval Procedure

Because it might take a long time for an IVF program to establish a high success rate based on the number of live babies born, it would be reasonable for a relatively new program to report the anticipated birthrate per egg retrieval procedure performed. This could be defined as the number of clinical pregnancies per egg retrieval that have progressed beyond the 12th week of gestation plus the number of live, healthy births that have occurred per egg retrieval. The acceptability of this statistic is based on the fact that once a pregnancy has proceeded beyond the 12th week it is highly unlikely to miscarry spontaneously.

In order to compute a mean anticipated egg retrieval procedure for the United States, several factors must be taken into account. First, it should be kept in mind that the mean IVF clinical pregnancy rate per egg retrieval is about 23%. Now, if 10% of these pregnancies are lost through early miscarriage and another 3 or 4% are miscarried after the 12th week, the probable birthrate per egg retrieval would be derived by reducing the 23% clinical pregnancy rate by about one-seventh (or 14%), so a guesstimate of the anticipated birthrate would be about 20% per egg retrieval. Therefore, when investigating a new program that can offer no other statistics, consumers should look for a probable birthrate per egg retrieval of about 20%.

If a program reports a lower rate, if it quotes figures in an ambiguous or confusing manner, or if it fails to provide evidence that the rate it is quoting is legitimate, then the couple who still choose that program have done so for reasons other than a rational expectation of success.

We strongly recommend that when asking for success rates the researching couple should hold IVF programs accountable by requesting that they provide all statistics in written form. It is also helpful to ask for long-term statistics (two to three years' worth) to account for the turnover in key staff members that frequently occurs in many IVF programs.

COUPLES HAVE THE RIGHT TO EXPECT COMPETENT, CARING TREATMENT

Another barometer of an IVF clinic's success is the way it treats its patients. A reputable IVF program should help each couple establish rational expectations right from the beginning and then follow through with a professional, understanding, organized program that meets the needs of both partners.

Consumers should look for a program that says, in effect: "We cannot guarantee that you will get pregnant, but we can promise you professionalism, the highest quality of care and expertise, a reasonable chance of getting pregnant, and that you will be dealt with all along the line with courtesy, understanding, and compassion."

Couples should look for a dedicated, committed team trained to deal with the emotional consequences of an IVF procedure and should avoid a program that is so preoccupied with the technical side of IVF that it loses sight of the human aspects of the procedure. No couple should feel they have to settle for a program that offers poor support and lacks compassion because they have nowhere else to go.

The morale and enthusiasm of the staff are good indicators of the kind of treatment the couple can expect. Morale in clinics that consistently report pregnancies is likely to be higher because staff members feel that they are part of a successful program. One program reinforces this enthusiasm by contacting patients who have a positive pregnancy test on a speakerphone. This enables everyone on the staff to share the joy and excitement with the couple.

Consumers might want to look for a program that offers a professional counselor to deal with both partners' emotional needs. The counselor, who is pivotal to any IVF program, usually acts as a buffer between the couple and the clinical team. Counseling can help a couple become positively involved in an IVF program and also steer them away from false hopes.

It is wise to inquire about the size of the staff and verify that the program has enough people to respond to the couple's needs at all times. No one wants to have to reschedule an egg retrieval because the doctor in an understaffed clinic was called away unexpectedly.

Care and caring go together in the truly successful IVF program. If the perception of caring truly indicates a successful program, then the program that elicited the following comment from this woman (who adopted a baby after her IVF pregnancy ended in miscarriage) must indeed be successful:

I don't think I'd try IVF again in the very near future because I have a six-week-old at home, but thanks to the staff I have a positive attitude and outlook about IVF and would seriously consider trying again later.

WHAT IS THE BEST WAY TO GET INFORMATION?

The most rational approach to assessing the IVF situation is by first becoming aware of the facts and statistics, asking pertinent questions according to one's own needs, and then actually visiting the site. A reputable program should be willing to answer questions and give the couple access to the facility.

When seeking information about a program the couple should look for staff who are willing to take the time to talk and to respond to questions frankly and openly. Some consumer-oriented staffs will even send literature about the program on request, as well as copies of articles from accredited professional journals, videotapes, stories about the program from newspapers and magazines, and sometimes names of previous patients who are willing to discuss their experiences.

If the clinic does not volunteer information, the couple may have to be assertive. At a minimum, they should expect to receive literature about how the program operates. The lack of such information for potential patients is a sign of poor organization. The couple should be wary of any program that refuses to provide information and statistics in writing or insists they come into the office. If the couple feel that they have to pry answers from an evasive staff, they might want to think twice about that program.

Preliminary Information Can Be Obtained by Telephone

The only way to ferret out success rates is by talking directly to someone at the clinic. We recommend that before calling a prospective program the couple should reread "How Is Pregnancy Defined?" and "How Should a Reported Pregnancy Rate Be Interpreted?" in this chapter. Then they should be prepared to ask the following questions:

1. How long has your program been established?
2. How many patients have you treated?
3. How many babies have been born?
4. How many egg retrievals have you performed?
5. How many embryo transfers have you done?
6. How many clinical pregnancies per egg-retrieval procedure have you recorded?
7. What is your miscarriage rate?
8. For established programs: How many deliveries per egg-retrieval procedure have you reported (birthrate per egg-retrieval procedure)?
9. For new programs: How many deliveries plus ongoing pregnancies that have proceeded beyond the 12th week have you experienced per egg retrieval (anticipated birthrate per egg-retrieval procedure)?
10. Do you have an embryo cryopreservation program? If so, how successful has it been? (Apply questions 1 through 9 to this issue.)

11. Do you offer ovum donation and IVF surrogacy (when applicable) in your program? (Apply questions 1 through 9 to this query as well.)

12. Do you turn away women over a certain age? If so, what age? (This is to be asked if age is a concern for the couple.)

13. Does your program perform intracytoplasmic sperm injection (ICSI) for the treatment of severe male infertility? How many procedures have been performed, and how many ongoing pregnancies and/or births have occurred?

It is the anticipated birthrate statistic that we consider to be the fairest and most helpful because it allows the newer clinics to provide an idea of the probability of having a live birth after IVF is done in their setting. Some consumers might give the benefit of the doubt, at least at first, to new clinics. However, if a program has not been in existence long enough to achieve any pregnancies, the staff should be forthright enough to explain that this is why they have no other statistics to offer.

In order to form the most rational expectations about each program, the couple should attempt to learn how the prognostic indicators for IVF (see Chapters 3 and 8) might impact on their personal chance of pregnancy in each particular program.

One way to do this would be to direct the conversation to their personal situation after having obtained general statistics about the program. The couple might first offer some information about themselves, including their ages, how long they have been infertile, what has been diagnosed as the cause of their infertility, the status of the man's fertility, and previous surgeries the woman may have undergone to correct her infertility. They should also be willing to supply other information the staff may request in order to become more familiar with the case.

Then the couple might ask:

1. In your program, what would be our chances per egg retrieval of conceiving a clinical pregnancy?

2. What would you say are our chances of actually having a baby after undergoing a single egg-retrieval procedure?

Once again, couples should request the answers to the questions in writing. Thereupon, after narrowing down the list of prospective clinics to those that responded most satisfactorily to these questions, the couple are ready for the next step—the pre-enrollment interview.

A Pre-Enrollment Interview Is Worth the Time and Expense

Just as few people would select a college without first visiting its campus, consumers also should visit each prospective program if at all possible. A program that refuses to grant a pre-enrollment interview should be dropped from further consideration.

A pre-enrollment interview will give the couple a chance to meet some of the staff and see what kind of people they will be dealing with. Is there an air of camaraderie, or do the staff seem disgruntled and unhappy? If the staff obviously regard their positions as nine-to-five drudgery, the couple most likely are in the wrong place.

Is the office comfortable and attractive? Does it create a relaxed, pleasant atmosphere? Does the program provide audiovisual equipment on which patients can watch informational tapes about IVF procedures? Of course audiovisual equipment is not required in order for a woman to get pregnant, but its availability indicates that the clinic cares enough to keep both partners informed and comfortable. A program that offers such amenities is one that cares for the emotional as well as physical needs of its patients.

The couple should be sure to meet the clinic coordinator during their visit because he or she is the person the couple will deal with daily. They should be sure the coordinator is in control of the program on a daily basis and will be congenial to work with.

If it is not possible to meet the doctor during the pre-enrollment interview, the couple might investigate how the doctor is viewed outside the clinic. Does he or she get along well with people? In vitro fertilization is a popular topic for discussion these days, and many people have strong opinions about the doctors who practice this specialty. The couple may be surprised at how easy it is to get that information.

If a pre-enrollment interview cannot be arranged, other approaches can be used to gather more information about a specific program. Phone calls to previous patients will be invaluable. The chapter of an infertility support group in the city where the program is located probably would be willing to help. The couple might even retain someone living near the clinic to conduct research for them. Perhaps the couple's own doctor knows a local physician who can provide information. The couple may even decide to arbitrarily telephone some OB-GYNs who practice in that community and ask them about the program.

While such research about a program can be helpful, in most cases nothing can really replace the information gained during a site inspection. A pre-enrollment trip is well worth the time and expense.

Consumers should expect to do a lot of homework when searching for an IVF program. Unfortunately, we do not believe that it will get any easier in the near future. As an IVF father told us:

> We have a library at home of all kinds of clippings, and virtually every book, magazine, and periodical you can imagine about IVF. My wife did a tremendous amount of research on which clinics were having the greatest success rate, what kinds of procedures were being used, what the latest technology was. Her training as a nurse certainly gave her a better handle on those strange-sounding hormones that are used as part of the process. Really, it was a matter of doing a lot of research for us before we were able to locate the right IVF program.

Helpful as it would be when selecting an IVF program, it is not necessary for every couple to have an RN in the family to use the guidelines suggested in this chapter. When consumers know what to look for and what questions to ask, they will be prepared to make an informed choice—a decision that should always be based on rational expectations, not false hopes.

12

GIFT AND OTHER ALTERNATIVES TO IVF

This chapter outlines some of the therapeutic gamete-related technologies available to the infertile couple. The term *therapeutic gamete-related technologies* refers to those procedures that involve enhancement, insemination, or transfer of eggs and/or sperm (gametes) into the woman's uterus, fallopian tubes, or peritoneal cavity in the hope that *in vivo* (inside the body) fertilization and the subsequent birth of one or more healthy babies will follow. In contrast, IVF involves fertilization in the laboratory and transfer into the uterus of an embryo or embryos rather than gametes.

We will recommend when these technologies should be considered in place of IVF and when IVF would be the best alternative. Again, we believe that a couple would be best advised to first consider the least invasive and/or least sophisticated gamete-therapeutic procedure that would meet their needs. For example, if they have the option of undergoing artificial insemination, or in cases where a diagnostic or therapeutic laparoscopy is scheduled, gamete intrafallopian tube transfer (GIFT) could be performed at the same time, provided that the woman's fallopian tubes are normal. This would provide the couple a chance of conceiving without any significant additional cost or risk.

ARTIFICIAL INSEMINATION

The procedures mentioned in this section are directed mostly but not exclusively to situations in which infertility is due to problems other than female organic pelvic disease and male-factor infertility. Indications for artificial insemination include cervical mucus insufficiency unrelated to sperm antibodies in the woman's secretions, unexplained infertility, and donor-sperm insemination. It is relatively contraindicated in situations of male-factor infertility, female immunologic infertility (due to sperm antibodies), tubal disease, and chronic pelvic adhesions. Couples for whom artificial insemination is indicated might consider the following alternatives before electing to undergo IVF. IVF would be performed if insemination procedures fail to achieve a pregnancy in spite of repeated attempts.

Intrauterine Insemination (IUI)

Intrauterine insemination (IUI), the injection of sperm into the uterus by means of a catheter directed through the cervix, has been practiced for many years. The premise of this procedure is that sperm can reach and fertilize the egg more easily if they are placed directly into the uterine cavity. In addition, in cases where the cervical mucus is poor or hostile to sperm, intrauterine insemination circumvents these problems because it bypasses the cervix.

In the early 1960s, physicians were attempting to enhance the chances of pregnancy occurring by injecting a small quantity of raw, untreated semen (sperm plus seminal plasma) directly into the uterus at the time of expected ovulation. However, when more than 0.2 ml of semen was inseminated directly into the uterus, a serious, sometimes life-endangering shocklike reaction often occurred. It was subsequently determined that the reason for this reaction was that the seminal plasma component of semen is rich in *prostaglandins.* When introduced directly into the uterus in large amounts, prostaglandins are capable of inducing serious and often life-endangering complications. However, the practice of restricting artificial insemination to less than 0.2 ml of semen largely prevents this threat. (Women are protected against the reaction during intercourse because the semen pools in the vagina; the sperm are then safely filtered through the cervical mucus, thereby preventing seminal plasma from reaching the uterine cavity.)

The results of intrauterine insemination with semen were dismal: capacitation was hardly likely to be initiated, and the uterus and fallopian tubes reacted defensively to the introduction of the seminal plasma (a foreign substance). The reason capacitation cannot be initiated properly in such a situation is because the seminal plasma contains anticapacitation factors that inhibit the entire process. Moreover, injecting uncapacitated sperm directly into the uterus preempts the important role that cervical mucus plays in initiating capacitation. Accordingly, any sperm that reach the waiting egg within the fallopian tube are unlikely to have the capacity to fertilize it.

When research demonstrated that the seminal plasma rather than the sperm caused the problem, it rapidly became common practice to wash the sperm by centrifugation and suspension, thus separating them from the seminal plasma. Washing the sperm offered three advantages: (1) it eliminated the risk to the woman of a prostaglandin reaction; (2) it got rid of the antimotility factors that are present in the seminal fluid and inhibit normal passage of sperm through the woman's reproductive tract; and (3) it did away with the anticapacitation factors in the seminal plasma that inhibit the proper initiation of sperm capacitation. Although the capacitation reaction was more likely to be initiated with washed sperm, it remained theoretically compromised because washing still did not compensate for the vital role the cervical mucus plays in nature.

Many women, however, did get pregnant after intrauterine insemination with washed sperm. In addition, reasonably good results were reported for selected cases of male subfertility and unexplained infertility. Why? Although it is believed that sperm must pass through the cervical mucus, the uterus, and the fallopian tubes to achieve capacitation, sperm apparently can also be capacitated, although less effectively, by coming in contact only with the secretions of the uterus and fallopian tubes.

In 1978, Sandra Allenson, a nurse-coordinator in one of our first IVF programs, suggested that because capacitation had to be initiated in preparation for IVF, it might be appropriate to consider initiating it in the sperm to be inseminated directly into the uterus. Ms. Allenson thought this would enhance the ability of the sperm to complete the capacitation reaction in the woman's body and to fertilize the awaiting egg or eggs. The process as applied to intrauterine insemination would involve, therefore, washing the sperm and separating them from the seminal plasma by centrifugation, followed by the laboratory procedure that is used to initiate capacitation in IVF (see "The Laboratory's Role in IVF" in Chapter 6). Thereupon, the washed and capacitated sperm would be injected into the uterus of a woman who had previously undergone controlled ovarian hyperstimulation with hMG/FSH. A similar technique can be done without prior COH, but the odds of pregnancy are much lower.

This approach to intrauterine insemination with COH resulted in a 37% pregnancy rate within three cycles of treatment in cases where the procedure was clearly the best option. At the time this book was written, many women had conceived in our programs through intrauterine insemination.

Since introducing intrauterine insemination in 1984, we have performed more than 2,000 procedures in women under 40 who had no evidence of organic pelvic disease. Our observations can be summarized, with two exceptions, as follows:

1. The best results with intrauterine insemination are obtained when hMG/FSH are used to induce ovulation. The two exceptions to this are artificial insemination with donor sperm and cases of cervical mucus hostility; insemination during natural cycles in normally ovulating women produces equally good results in both situations. The success rate in these categories is around 20% per cycle of treatment.

2. When clomiphene citrate is used for COH, the success rate is about one-half that achieved through the use of hMG/FSH.

3. The best results with intrauterine insemination are obtained in cases of unexplained infertility, abnormal ovulation, cervical mucus hostility unrelated to sperm antibodies in the woman, and when donor sperm is used.

4. The poorest results with intrauterine insemination are seen in cases of moderate or severe male infertility (success rates are less than 7% per cycle) and where IUI is performed in natural (unstimulated) cycles. The exceptions to IUI in unstimulated cycles were cases of donor insemination and cervical mucus hostility due to female sperm antibodies.

5. Normally ovulating women who have mild pelvic endometriosis (with no pelvic adhesions) also have reduced pregnancy rates (about 8% per cycle following COH). This underscores the fact that even in the absence of pelvic adhesions or damage to the fallopian tubes and ovaries, endometriosis creates an unfavorable pelvic environment that compromises fertilization of the woman's eggs after ovulation and during passage of the egg or eggs from the ovary to the fallopian tube.

Pregnancy rates following IUI performed in women over the age of 40 are about one-half that achieved in younger women. This is probably largely due to the adverse effect of age on egg and embryo quality. We have also observed that women over 40 rarely achieve viable pregnancy following clomiphene therapy. They do ovulate, but they usually have poor uterine linings and hostile cervical mucus.

We therefore believe that women over 40 who undergo IUI should receive hMG/FSH, not clomiphene, because of the urgency brought about by age. The best option for such couples is to make no more than three or four attempts at IUI with washed and capacitated sperm, whereupon IVF should be undertaken before it is too late. In other words, women over 40 should seriously consider the age factor in choosing between IUI and IVF.

Intrauterine insemination with washed and capacitated sperm can be performed at about 20 to 25% the cost of an IVF procedure. Moreover, intrauterine insemination is often more likely than IVF to be considered reimbursable by many insurance companies.

Intravaginal Insemination (IVI) with Partner's Semen

Intravaginal insemination (IVI) using the partner's semen involves the injection of semen into the vagina in proximity to the cervix rather than into the uterus, as is the case with IUI. Intravaginal insemination is most often employed to assist a woman with a subfertile partner to conceive naturally at the time of ovulation. However, IVI usually offers no advantage over normal ejaculation that occurs during intercourse. The only cases when IVI might be advantageous would be certain forms of male impotence in which the man cannot produce semen with intercourse.

Artificial Insemination by Donor (AID)

Artificial insemination by donor (AID) is the most common form of insemination in cases in which donor sperm are required because the woman's partner is infertile. Artificial insemination by donor can be done via IVI or IUI.

The use of cryopreserved donor sperm is the safest method of performing donor insemination. Given the significant risk of AIDS in the modern setting, there is hardly any justification for using fresh donor sperm. After the donor has been tested for AIDS, the sperm are cryopreserved for at least six months, following which the donor is retested for AIDS. This double testing is necessary because a man may not register positive for up to six months after infection by the AIDS virus. If the AIDS virus is not present, the likelihood that the original specimen is infected is remote. The specimen is then released.

The emergence of new tests that can accurately detect the AIDS virus without the need to wait for the development of antibodies could change the situation and might in the future again permit the use of fresh sperm.

GAMETE INTRAFALLOPIAN TRANSFER (GIFT)

In 1984, Dr. Ricardo Asch introduced a therapeutic gamete-related technique that has gained widespread popularity in the United States. It involves the injection of one or more eggs mixed with washed, capacitated, and incubated sperm directly into the fallopian tubes. Dr. Asch is believed to have devised the acronym GIFT—gamete intrafallopian transfer—in order to promote the concept that GIFT gives the gift of life.

Gamete intrafallopian tube transfer is usually done through laparoscopy. It can also be performed *transcervically* by the introduction of a catheter through the cervix into one of the fallopian tubes, where the sperm and eggs are discharged. This latter approach, however, although having yielded pregnancies in the past, has produced very disappointing results and has not gained widespread popularity. Accordingly, for all practical purposes the performance of GIFT requires laparoscopy.

The woman is usually stimulated with fertility drugs in order to achieve superovulation before the eggs are harvested for a GIFT procedure. The eggs can be removed through vaginal ultrasound-guided needle-aspiration prior to administering general anesthesia or at the time of laparoscopy through the introduction of a needle via the abdominal wall. The former approach has the distinct advantage of allowing cancellation of the laparoscopy if no eggs are retrieved. The eggs are then mixed with sperm that have been previously washed and capacitated in the laboratory, and then both the eggs and sperm are loaded into a fine catheter.

If GIFT is performed during laparoscopy, the physician injects the eggs and sperm into the fallopian tubes under direct vision through the laparoscope. In this case of minilaparotomy, the physician gently moves the ends of one or both fallopian tubes outside the abdomen through the incision, injects the sperm and eggs directly into the tubes, and then carefully returns them to the abdominal cavity before closing the abdomen.

The advantage that GIFT holds over intrauterine insemination is that GIFT ensures that the eggs and sperm arrive simultaneously at the point in the fallopian tubes where fertilization would normally occur. By placing the eggs and sperm together in the outer third of the fallopian tubes, GIFT eliminates any concern regarding the ability of the fimbriae ends of the fallopian tubes to pick up or receive the eggs at the time of ovulation. In effect, GIFT substitutes incubation in the body for incubation in the petri dish prior to fertilization.

Because GIFT as currently performed requires laparoscopy, it is also significantly more expensive and physically demanding than most other techniques described in this chapter. Its cost approaches that of IVF.

While GIFT, in contrast to IVF, does not require the relatively costly fertilization process in the laboratory, it does require laparoscopy and general anesthesia. This negates any potential cost savings over IVF.

Another major disadvantage of GIFT is that it lacks IVF's diagnostic capacity, which enables the physician to see in the laboratory whether the woman's egg can be fertilized by her partner's sperm. As we explained earlier, male infertility and many cases of unexplained infertility are related to the fact that the sperm are incapable of fertilizing the eggs or that the eggs are unfertilizable. If GIFT is unsuccessful, the physician has no way of knowing whether it failed because the eggs could not be fertilized or whether other factors were responsible.

When sperm dysfunction is the primary problem, it makes sense to use only the best available tool for the job. The intensive care provided in the IVF laboratory maximizes the chances for fertilization. GIFT, in contrast, simply places a large number of sperm in the fallopian tube near an otherwise unprepared egg; hence, fertilization is much more of a hit-or-miss situation. Accordingly, we believe that GIFT, with a few exceptions, is not a good choice for treatment of unexplained or male infertility unless the woman requires a laparoscopy for reasons other than infertility treatment and GIFT can be performed concomitantly.

Ectopic Pregnancies with GIFT

It was initially believed that injection of the male and female gametes into the fallopian tubes during GIFT might increase the risk of tubal pregnancies. However, statistics have not borne out this concern in cases in which the tubes are apparently normal in configuration and where no other pelvic disease is present.

A significant increase in the incidence of ectopic pregnancies has, however, been reported when GIFT is performed in abnormal fallopian tubes. Such pregnancies might occur even though tubal reconstructive surgery appears to have restored the patency as well as the outward appearance of fallopian tubes previously distorted by disease. Although the outward appearance of fallopian tubes may suggest they are normal, there is currently no reliable method of determining whether their internal integrity has been restored. Diagnostic procedures such as hysterosalpingogram, or injection of dye, may reveal that the tubes are open; but such examinations by no means assess whether the inner lining of the tube has been partially damaged or if the wall of the tube might be less mobile than desired. The future application of falloposcopy as a diagnostic tool might help assess the intratubal environment and better evaluate patients as candidates for the GIFT procedure.

Undetected defects in the interior of the fallopian tubes can lead to devastating consequences from GIFT. For example, damage to the interior of a fallopian tube from disease might inhibit normal physiologic function and/or peristaltic movements. Thus, the embryo might not be propelled toward the uterus in a timely manner. (Figure 2-9 depicts the day-by-day timeline for normal transport of the zygotes and embryos through the fallopian tube to the uterus.) If nature's schedule is delayed because of sluggish peristaltic movements, the embryo might hatch in the fallopian tube and grow into its wall, thus creating an ectopic pregnancy.

Therefore, because if left undiagnosed ectopic pregnancies are often life-endangering, we firmly believe that GIFT should be reserved for cases in which there is no evidence of previous or existing tubal disease even though the tubes might appear to be normal or to have been restored to normalcy through surgery.

Pregnancy Rates with GIFT

The birthrates reported with GIFT performed on women under 40 range from 25 to 35%. These statistics are significantly higher than the average national statistics reported for IVF, which has led many people to believe that GIFT is superior to IVF. In addition, the relatively low success rates reported with IVF in women over 40 has led some physicians to advocate performing GIFT rather than IVF in these women.

It is important to keep in mind, however, that GIFT is a relatively easy procedure that does not require anywhere near the degree of sophisticated technical expertise in the laboratory that is required in IVF. Even the poorest IVF programs report satisfactory results with GIFT, while only the best IVF programs report high success rates with IVF. The results from GIFT must be viewed against this backdrop.

It is also true that good IVF programs are able to produce far better than national average success rates. Furthermore, IVF is much less invasive because it does not require the performance of laparoscopy. For this reason, it is the preferred approach in most programs that report good IVF results. It should be noted that there is no evidence to support the thesis that the performance of GIFT in a woman over 40 affords a better chance of pregnancy than IVF in a good program.

Finally, GIFT does not afford the opportunity to use leftover eggs (those not transferred to the fallopian tubes). In contrast, IVF permits leftover embryos to be cryopreserved.

ZYGOTE INTRAFALLOPIAN TRANSFER OR TUBAL EMBRYO TRANSFER (ZIFT/TET)

Another option for achieving pregnancy in cases where infertility is unrelated to female organic pelvic disease involves the transfer of one or more zygotes (fertilized eggs) or embryos directly into the woman's fallopian tubes during laparoscopy. As with routine IVF, this procedure requires an initial egg retrieval and fertilization of the eggs in the laboratory. Thereupon the zygotes or embryos are loaded into a thin catheter and injected into the outer third of one or both fallopian tubes during laparoscopy. The egg retrieval is performed through transvaginal needle-aspiration, and one or two days later zygotes are transferred to one or both fallopian tubes during the laparoscopic procedure.

In the past, proponents of ZIFT/TET argued that enabling the embryo to reach the uterus via its natural route (the fallopian tube) rather than by embryo transfer through the cervix increases the likelihood of implantation and a successful pregnancy. They also contended that ZIFT/TET would allow the embryos to travel down the fallopian tube on their own, and so reach the uterus at the appropriate stage of cleavage (about five days after transfer) when the uterus is optimally prepared, while IVF delivers an embryo directly into the uterus two or three days earlier. Accordingly, it was argued that ZIFT/TET was more advantageous, especially for the older woman, whom IVF offers a lower success rate.

It must be emphasized that the studies that previously reported encouraging and even superior results with ZIFT/TET were all poorly controlled. More recently, a number of well conducted studies have confirmed that the pregnancy rate per embryo transferred with ZIFT/TET is the same as that reported for IVF. Accordingly, there is hardly any justifiable indication for the performance of ZIFT/TET in preference to IVF.

NATURAL CYCLE IVF

Natural cycle IVF involves accessing one and sometimes two follicles that might develop in a woman during a normal cycle for the purpose of fertilizing the eggs in vitro and transferring them to the uterus. Several advantages of this method are often cited: (1) The woman's cycle is not affected by fertility drugs, so she should have an optimum environment into which to place the embryos; (2) normal unstimulated cycle IVF obviates the use of fertility drugs, which have certain emotional side effects and other rumored but unproved side effects, such as ovarian cancer; and (3) the cost is considerably less than for IVF.

But an objective look at these arguments finds considerable evidence to the contrary. First, the amount of monitoring in a natural cycle significantly exceeds that which has to be done in a planned IVF cycle. The laboratory is still necessary, the eggs must be harvested, the sperm prepared, the embryo or embryos transferred, and an increased amount of blood testing is required in order to accurately monitor the woman's progress. Natural-cycle patients must be closely monitored at all times—including evenings and weekends—and the unpredictability of follicle maturation without the trigger of hCG often results in nighttime or weekend egg retrievals, which drive up the cost because of the need for increased staff.

Therefore, it can be concluded that with the exception of the cost of fertility drugs, the performance of natural-cycle IVF does little to lower the overall cost of the procedure. The success rates from natural-cycle IVF are very much lower than from conventional IVF. Most programs doing natural-cycle IVF report no more than a 10 to 15% pregnancy rate per cycle with an anticipated birthrate of no more than 8 to 10% per cycle. Moreover, a woman stimulated with fertility drugs is likely to produce enough eggs so that some can be cryopreserved for later use, thus giving the couple an additional opportunity to achieve a pregnancy in a subsequent cycle; this, of course, is not an option during a natural cycle.

If natural-cycle IVF has any place at all, it would be with a young woman who has absolutely normal ovulation and hormonal balance and where the only cause of infertility is tubal occlusion. The woman would preferably be under 30, definitely under 35, because of the subtle abnormalities of the hormonal balance in older women that tend to make the endometrial environment less favorable for implantation.

CHAPTER

13

THIRD-PARTY PARENTING:

OTHER OPTIONS FOR COUPLES WITH INTRACTABLE INFERTILITY

For many couples who are unable to achieve pregnancy through conventional treatments, third-party parenting offers tremendous hope for success. Third-party parenting is a collective term for egg donation, embryo adoption, gestational surrogacy, donor insemination, and adoption of a child. These procedures are options for the infertile couple to consider when the woman, for some reason, cannot produce healthy eggs or the proper gestational environment for a pregnancy, or when the man cannot produce healthy sperm.

Only a few years ago, women who did not have a healthy uterus and those who could not produce healthy eggs had the lowest chance of having their own baby. Now, quite paradoxically, through the advent of egg donation and IVF/surrogacy (IVF third-party parenting), these women have the greatest chance by far of conceiving, greater than with any other cause of infertility.

Some women are born without a uterus, while others undergo surgical removal of the uterus in later life. Sometimes uterine disease renders the woman incapable of bearing a child, and, in a minority of cases, chronic ill health, such as severe diabetes, makes pregnancy inadvisable. For these couples, the option exists of having another woman—a third party or surrogate—bear a child for them. Surrogate parenting can be divided into two categories: *classic surrogacy* and *IVF surrogacy.*

In classic surrogacy, a healthy young woman (usually under 35) agrees with an infertile couple to be artificially inseminated with the male partner's sperm, carry the baby to term, and then turn the baby over to the couple shortly after birth.

Classic surrogacy has brightened the lives of many desperate infertile couples, but it also brings with it many ethical, moral, and medico-legal dilemmas. There is no getting around the fact that because the classic surrogate provides both the egg and the womb, she is biologically the child's mother. This is the primary cause of surrogates' last-minute decisions not to give up the child. Who can ignore the intense media coverage that often erupts when a surrogate decides against giving up the baby to the infertile couple? Situations like this cause wrenching emotional turmoil for the parents, for the surrogate, and (sooner or later) for the child. Classic surrogacy currently is, nevertheless, the most widely employed method of surrogate parenting.

Since we do not offer classic surrogacy in our programs at the present time, we will not discuss it further here. However, we will examine egg donation, embryo adoption, and IVF surrogacy in detail in the following sections.

EGG DONATION

For some infertile women, disease and/or the onset of ovarian failure precludes their ability to produce a fertilizable egg. But since they have a healthy uterus and are otherwise able to bear a child, egg (or ovum) donation offers a realistic opportunity for pregnancy. Egg donation involves retrieving eggs from one woman (the donor), fertilizing them in the laboratory, and transferring the resulting embryos into the uterus of the recipient (the woman partner in the infertile couple), who will carry the baby to term. The donor is stimulated with fertility drugs, and the eggs are fertilized with the partner's sperm.

Around the age of 45 a woman has few other choices if she wishes to carry a baby herself. This is due to the impact of age on egg quality. Women between 40 and 45 or those nearing an early menopause are often able to choose whether they want to try getting pregnant with donor eggs or their own. This highly personal choice should be considered in light of both the financial and emotional costs.

The couple must expect that the financial demands for these high-tech procedures will be in the conventional IVF price range. In addition, the woman 40 to 45 years of age who opts to use her own eggs has at best a 12 to 15% chance per egg retrieval procedure of having a baby through IVF. Accordingly, in order to achieve pregnancy using her own eggs, she and her partner must realistically be able to afford a number of treatment cycles.

Conversely, when a 40- to 45-year-old woman decides to use donor eggs (provided she has a normal uterus, a fertile partner, and uses eggs from a woman under 35 years), she can anticipate a birthrate per cycle of IVF that exceeds the optimum reported for young women. In other words, the effect of age on outcome with IVF is largely negated through the use of donor eggs. This is because the donor eggs will come from a younger woman and will therefore be healthy. In addition, the newer methods for preparing the recipient's uterine lining optimize the chance of healthy implantation.

Many times more eggs are retrieved from a young donor than are needed for any one pregnancy attempt. In this case, we fertilize the extra eggs and freeze many healthy young embryos for a future attempt at pregnancy. Use of cryopreserved embryos in subsequent "frozen-embryo transfer cycles" (performed in our setting) gives such egg-donation patients approximately another 30% chance of having a baby following one egg harvest. A further benefit of egg donation is a reduced risk of miscarriage, because of the quality of the younger woman's eggs. In addition, because the eggs will usually come from a relatively young and healthy donor, there is rarely a need for amniocentesis or chorionic villus sampling in order to diagnose chromosome disorders that produce birth defects.

On the emotional side, the long-term quest for pregnancy is stressful at any age. After the age of 40 it takes on the added stress of the relentlessly ticking biological clock. The couple must assess whether they can withstand the many possible disappointments on the road to child bearing. How important is it now that the child be genetically theirs? Is it more important to them at this point to achieve a pregnancy with donor eggs and get on with their lives? Potential parents have to answer these questions for themselves.

Candidates for egg donation/IVF should meet one or more of the following criteria:

1. Ovarian resistance to stimulation with fertility drugs.
Despite repeated attempts, the ovaries fail to produce several eggs when
stimulated with maximum doses of fertility drugs.

**2. Poor fertilization of the woman's own eggs in prior IVF
attempts, in spite of good-quality sperm.**

**3. Failure to achieve a viable pregnancy following repeated
attempts at IVF, GIFT, or tubal embryo transfer (TET).** In spite
of repeatedly transferring four or more embryos or gametes to the uterus
or fallopian tubes with IVF or GIFT, pregnancy fails to occur. However, in
such cases it is essential to first rule out any factors that might be
compromising healthy implantation.

**4. Absence of ovarian function due to surgery, radiation, or
chemotherapy for malignant diseases.**

5. Premature menopause. Women who undergo menopause under
the age of 40 and whose uterus is capable of responding to hormonal
treatment are ideal candidates for egg donation/IVF.

6. Menopause. Women over 40 whose ovarian function has ceased
as a result of surgery, infection, or endometriosis are good candidates
provided there is no uterine factor inhibiting implantation.

**7. The presence of genetic disorders that have a high
likelihood of being transmitted via the woman's eggs to the
offspring.** Some of these disorders cannot be readily diagnosed through
amniocentesis or chorionic villus sampling; in such cases, egg
donation/IVF may be indicated.

**8. Raised levels of FSH during first three days of the
menstrual cycle.** This too might indicate a poor response to ovarian
stimulation with fertility drugs.

Beginning with selection of a donor, the following section outlines what
an infertile couple might experience when entering an embryo adoption
program. This is not necessarily the only way it should be done, though the
section does include a number of procedures we consider to be basic to
any such program.

Selecting the Donor

Our program solicits the services of a reputable ovum donor/surrogacy
agency with access to many donors and surrogates. This provides a wide
choice for couples with diverse needs. Other programs compile lists of

available donors who have been screened genetically and clinically as well as thoroughly evaluated for hepatitis, HIV, and other sexually transmitted diseases. *We recommend that the donor be anonymous in most cases.* However, a couple could make special arrangements to use a donor they know provided that this is clearly defined and agreed upon at the outset. Every attempt is made to match the donor with the recipient relative to the recipients' request. Issues such as physical characteristics, race, ethnic background, religion, etc. are all disclosed.

Screening the Donor and Recipient

In addition to undergoing the customary physical examination and tests, the donor also undergoes a psychological evaluation in our setting. She will also have blood drawn on the first or second day of a natural menstrual cycle for the measurement of FSH and estradiol. In addition, she will have blood tests for AIDS, hepatitis, and other sexually transmitted diseases.

Both the woman who is to receive the embryos (recipient) and her male partner will undergo thorough testing, including a careful clinical evaluation as well as the following tests:

1. A cervical culture to evaluate the presence of organisms, such as chlamydia or ureaplasma, that might interfere with a successful outcome.
2. Hysteroscopic examination of the woman's uterus for lessions that might interfere with implantation (see Chapter 8).
3. A psychological assessment of the recipient and her partner (when applicable) with a psychological counselor.
4. Analysis of the male partner's semen.

Finally, the couple will visit with the clinical coordinator, who will outline the exact process step by step. Once all the evaluations have been completed, they will select a date to begin treatment.

Stimulating and Monitoring the Donor

In order to stimulate ovulation of enough eggs to increase the chances of a viable pregnancy, the donor will be treated with gonadotropins. But first, she will be asked to use barrier contraception or to abstain from sexual intercourse in the cycle immediately prior to stimulation. Approximately seven days after ovulation occurs (as assessed by a BBT chart or a urine home-ovulation test kit), GnRHa is administered daily to prepare the ovaries. With the onset of menstruation approximately 7 to 12 days later, the donor is given a blood test and baseline ultrasound examination to confirm that the ovaries are prepared and to exclude the presence of ovarian cysts. The decision is made then about when gonadotropin therapy should commence.

The donor's first day of gonadotropin injections is referred to as cycle day 2. On cycle day 9, the program would likely begin intensive daily monitoring by means of blood hormone measurements and ultrasound examinations. Usually, one to three additional days of gonadotropin therapy will be required. Once monitoring confirms that the donor's ovarian follicles have developed optimally, she will receive an injection of the ovulatory trigger hCG. Then, in order to capture the eggs prior to ovulation, they are harvested 36 hours after the hCG injection.

Synchronizing the Cycles of the Donor and the Recipient

It is absolutely critical that both women's cycles be synchronized as closely as possible so that the endometrial lining of the recipient's uterus can be prepared for implantation of the transferred embryos. The female hormones estrogen and subsequently progesterone will be given to the recipient to prepare the endometrium. When the recipient is menopausal and therefore her ovaries are inactive, preparatory hormonal therapy can, in such cases, be initiated without GnRHa. For recipients with residual ovarian function and who are therefore usually still menstruating, 7 to 12 days of GnRHa therapy is required in order to achieve ovarian desensitization prior to commencing the hormone injections. The duration of GnRHa therapy is adjusted to synchronize the recipient's cycle with that of the donor.

Building the Recipient's Uterine Lining with Hormonal Injections

The recipient receives estrogen orally, by skin patches, or by injection. In our programs we administer estradiol valerate by injection on Tuesdays and Fridays, and draw the recipient's blood on Mondays and Thursdays to measure estradiol concentrations in order to determine the subsequent hormonal dosage. She also undergoes ultrasound examinations at least weekly to evaluate the development of her uterus's endometrial lining. Two days prior to the expected day of embryo transfer, the recipient is given daily injections of progesterone to optimize endometrial development.

In the uncommon event of poor endometrial development, the couple will be given the choice of either having the donor's eggs harvested, fertilized, and frozen for transfer to the recipient's uterus in a subsequent cycle or canceling the procedure.

The Donor Undergoes Transvaginal Ultrasound-guided Egg Retrieval, Egg Fertilization, Embryo Culture, and Embryo Transfer

All of these procedures are described in detail in Chapters 6 and 7.

Management and Follow-up After the Embryo Transfer

The recipient continues to receive daily progesterone and biweekly estradiol valerate injections in order to sustain an optimal environment for implantation, and a pregnancy test is performed about 10 days after embryo transfer. A positive test indicates that implantation is underway. In such an event, the hormone injections will be continued for an additional six to eight weeks. In the interim, an ultrasound examination will be performed to definitively diagnose a clinical pregnancy. If the test is negative, all hormonal treatment is discontinued and menstruation will ensue within three to 10 days.

If the recipient does not conceive, she may have her frozen embryos, if any, thawed and transferred to her uterus in a subsequent cycle. If in spite of both the initial attempt and the subsequent transfer of thawed embryos the recipient does not conceive, she may schedule a new cycle of treatment.

Anticipated Success Rates with Egg Donation/IVF

Reported birthrates for egg donation/IVF have traditionally paralleled those achieved through conventional IVF (20 to 30% per embryo transfer procedure, depending on the center where the procedure is performed). However, now that in our programs we hormonally prepare the uterus of the woman who will be carrying the pregnancy with estradiol valerate injections prior to the embryo transfer, we have experienced birthrates of approximately 50% or greater per egg retrieval. This is much higher than the national average. But these statistics apply only when the eggs are derived from a woman under the age of 35 and the recipient has a healthy uterus. The birthrate at Pacific Fertility Center in San Francisco following the transfer of thawed embryos derived from donor eggs is about 30% per embryo transfer procedure.

EMBRYO ADOPTION

Embryo adoption refers to the situation in which a woman receives embryos to which she and her male partner have not contributed biologically. When both partners are infertile, both donor sperm and donor eggs must be used if the woman is to become pregnant. Previously, adoption of a child would have been such a couple's only option. Now, however, "prenatal embryo adoption" can be an alternative to adoption of a baby or child. We perform these adoptive procedures because we believe that apart from the fact that embryo adoption occurs far earlier than baby adoption, there is otherwise little difference between the two processes.

Donor embryos can come from several sources. For example, a woman who cannot produce her own eggs might choose to adopt one or more embryos from a donor and have them transferred into her uterus. An additional source of embryos would be couples who, finding they have more embryos than they wish to transfer after IVF, donate the extras to another couple.

IVF SURROGACY

IVF surrogacy involves the transfer of one or more embryos derived from the infertile woman's eggs and from sperm of her partner (or a sperm donor) into the uterus of a surrogate. In this case, the surrogate provides a host womb but does not contribute genetically to the baby. While ethical, moral, and medico-legal issues still apply, IVF surrogacy appears to have gained more social acceptance than classic surrogacy. We offer IVF surrogacy as an option in most of our programs.

Candidates for IVF/Surrogate Parenting

Candidates for IVF surrogacy can be divided into two groups: (1) women born without a uterus or who because of uterine surgery (hysterectomy) or disease are not capable of carrying a pregnancy to full term; and (2) women who have been advised against undertaking a pregnancy because of systemic illnesses, such as diabetes, heart disease, and hypertension, or certain malignant conditions.

As in preparation for other assisted reproductive techniques, the biological parents undergo a thorough clinical, psychological, and laboratory assessment prior to selecting a surrogate. The purpose is to exclude sexually transmitted diseases that might be carried to the surrogate at the time of embryo transfer. They are also counseled on issues faced by all IVF aspiring parents, such as the possibility of multiple births, ectopic pregnancy, and miscarriage.

All legal issues pertaining to custody and the rights of the biological parents and the surrogate should be discussed in detail and the appropriate consent forms completed following full disclosure. We recommend that the surrogate and biological parents get separate legal counsel to avoid the conflict of interest that would arise were one attorney to counsel both parties.

In order to protect medical and administrative staff from having to confront couples with a bill in cases where a cycle of treatment fails to result in a healthy pregnancy, we require payment for services in advance.

Selecting the Surrogate

Many infertile couples who qualify for IVF surrogate parenting solicit the assistance of empathic friends or family members to act as surrogates. Other couples seek surrogates by advertising in the media. Many couples with the necessary financial resources retain a surrogacy agency to find a suitable candidate. We direct our patients to a reputable surrogacy agency with access to many surrogates. Because the surrogate gives birth, it is rarely possible or even realistic for her to remain anonymous.

Screening the Surrogate

Once the surrogate has been selected, she will undergo thorough medical and psychological evaluations, including:

1. A cervical culture and/or DNA test to screen for infection with chlamydia, ureaplasma, gonococcus, and other infective organisms that might interfere with a successful outcome.

2. Blood tests (as appropriate) for HIV, hepatitis, and other sexually transmitted diseases. She will also have a blood test performed to ensure

that she is immune to the development of rubella (German measles) and will have a variety of blood-hormone tests, such as the measurement of plasma prolactin and thyroid-stimulating hormone (TSH).

Whether recruited from an agency, family members, or through personal solicitation, the surrogate should be carefully evaluated psychologically as well as physically. This is especially important in cases where a relatively young surrogate or family member is recruited. In such cases, it is important to ensure that the surrogate has not been subjected to any pressure or coercion.

The surrogate should also be counseled on issues faced by all IVF aspiring parents, such as multiple births. She should also visit with the clinical coordinator, who will outline the exact process step by step. She should be informed that she has full right of access to the clinic staff and that her concerns will be addressed promptly at all times. And she should be aware that if pregnancy occurs, she will be referred to an obstetrician for prenatal care and delivery.

In the event that a viable pregnancy is confirmed by ultrasound recognition of a fetal heartbeat (at the sixth week), there is a better than 90% chance that the pregnancy will proceed normally to term. Once the pregnancy has progressed beyond the 12th week, the chance of a healthy baby being born is upward of 97%. In our setting, we anticipate approximately a 50% birthrate every time embryos are transferred to a surrogate, provided the biological mother (the egg provider) is under 35 and the surrogate has a healthy uterus. The birthrate declines as the age of the egg provider advances beyond 35. It is important to note that there is no convincing evidence to suggest an increase in the incidence of spontaneous miscarriage or birth defects as a direct result of IVF surrogacy.

If the surrogate's blood pregnancy tests are negative, treatment with estrogen and progesterone is discontinued, and she can expect to menstruate within four to 10 days. In the event that the pregnancy test is positive, estrogen and progesterone therapy will continue for about six weeks.

After the evaluations and counseling of both the couple and the surrogate have been completed, the three of them will meet. And once all the evaluations have been completed, the couple will select a date to begin treatment.

Follicular Stimulation and Monitoring the Female Partner (Egg Provider)

The procedure used to stimulate the female partner of the infertile couple with fertility drugs and monitor her condition strongly resembles that used for an egg donor. In order to stimulate ovulation of enough eggs to increase the chances of a viable pregnancy, the female partner will be stimulated with hMG/FSH. Approximately seven days after ovulation occurs (as assessed by a BBT chart or a urine home-ovulation test kit), GnRHa is administered daily to prepare the ovaries. With the onset of menstruation approximately 7 to 12 days later, the female partner is given a blood test and baseline ultrasound examination to confirm that the ovaries are prepared and to exclude the presence of ovarian cysts. The decision is made then about when hMG/FSH therapy should commence.

The female partner's first day of hMG/FSH injections is referred to as cycle day 2. On cycle day 9, the program would likely begin intensive daily monitoring by means of blood hormone measurements and ultrasound examinations. Usually, one to three additional days of hMG/FSH therapy will be required. Once monitoring confirms that the female partner's ovarian follicles have developed optimally, she is given an injection of the ovulatory trigger hCG. Then, in order to capture the eggs prior to ovulation, they are harvested 36 hours after the hCG injection by transvaginal ultrasound needle-guided aspiration.

Synchronizing the Cycles of Surrogate and Aspiring Mother

The surrogate will receive estrogen orally, by skin patches, or by injections, and then progesterone to help prepare her uterine lining for implantation. As with preparing the recipient for IVF/ovum donation, we use biweekly estradiol valerate injections in our programs. GnRHa is administered for a period of 7 to 12 days in order to prepare the ovaries prior to administration of estradiol valerate. The duration of GnRHa therapy is adjusted to synchronize the cycle of the woman undergoing follicular stimulation with that of the surrogate. Once the prospective mother commences follicular stimulation, the surrogate will be given estradiol and progesterone injections while continuing GnRHa therapy.

Building the Surrogate's Uterine Lining with Hormonal Injections

In our programs, the surrogate receives estradiol valerate injections on Tuesdays and Fridays, and her blood is drawn on Mondays and Thursdays to measure estradiol concentrations so the physician can determine the subsequent hormonal dosage. She also undergoes ultrasound examinations 10 days to two weeks after the first estradiol valerate injection to evaluate development of her uterine lining. Approximately four days prior to the expected day of embryo transfer, the recipient is given daily injections of progesterone to optimize endometrial development. In the uncommon event of poor endometrial development, the couple will be given the choice of either having the aspiring mother's eggs harvested, fertilized, and frozen for transfer to a surrogate's uterus in a subsequent cycle, or canceling the procedure.

The Egg Provider (Aspiring Mother) Undergoes Transvaginal Ultrasound-Guided Egg Retrieval, Egg Fertilization, and Embryo Culture

All of these procedures are described in detail in Chapters 6 and 7.

Transferring the Embryos to the Surrogate's Uterus

Approximately 48 to 72 hours following egg retrieval, the embryos are transferred to the surrogate's uterus. She then lies perfectly still for approximately two hours to enhance the chances of implantation and is then discharged from the clinic.

Management and Follow-up After the Embryo Transfer

The surrogate will be given daily progesterone injections and biweekly estradiol valerate injections and/or suppositories in order to sustain an optimal environment for implantation, and approximately 10 days after the embryo transfer will undergo a pregnancy test. A positive test indicates that implantation is taking place. In such an event, the hormone injections will be continued for an additional four to six weeks. In the interim, an ultrasound examination will be performed to definitively diagnose a clinical pregnancy. If the test is negative, all hormonal treatment is discontinued, and menstruation will ensue within three to 10 days.

If the surrogate does not conceive, the aspiring mother may have her remaining embryos frozen, to be thawed and transferred to the uterus of another woman at a later date. If in spite of both the initial attempt and subsequent transfer of thawed embryos the surrogate does not conceive, the infertile couple may schedule a new cycle of treatment.

Anticipated Success Rates with IVF Surrogacy

Like the pregnancy rates for egg donation, the rates for surrogation have traditionally paralleled those achieved through conventional IVF.

However, in the case of surrogacy where the age of the egg provider cannot be controlled, success rates are influenced by the effect of age on egg and embryo quality. The implementation of a new method of preparing the uterus with estradiol valerate injections for embryo transfer has resulted in viable pregnancy rates greater than 60% per embryo transfer when the eggs are derived from women under 40 years of age. As with IVF/ovum donation, when four or more embryos derived from the eggs of a woman under 35 are transferred to a surrogate with a healthy uterus, we achieve viable pregnancy rates of greater than 60% per embryo transfer procedure and birthrates of nearly 50%.

TOWARD THE BIOETHICS OF IVF SURROGACY

The determination of ethical guidelines has not kept pace with the exploding growth and development in IVF. However, some leaders in the field are working together, sharing experiences and advice, in an attempt to formulate a code of ethics. We end this chapter with a suggestion made by Dr. William Andereck in 1993 in a presentation called "Ethical Issues in the New Reproductive Technologies." He cited what he calls the "two-out-of-three rule" that he has applied to gestational surrogacy:

> *The genetic combination of the male and the female provide two of the essential elements which, along with gestation, are necessary to produce a human being. The two-out-of-three rule basically looks at these three elements: the egg, the sperm, and the gestational component. If at all possible, I recommend that at least two of these three components be contributed by the intended parents. If they can only contribute one, by all means please try not to get the other two contributed by the same person.*

This is a good first step that can be applied to many of the situations discussed in this chapter.

CHAPTER
14

ETHICAL IMPLICATIONS OF FERTILITY TECHNOLOGY

What would George Orwell have said about new fertility techniques such as *cryopreserving* (freezing and storing in liquid nitrogen) eggs, sperm, and embryos for future use?

CRYOPRESERVATION AS AN OPTION: FROZEN VS. FRESH SEMEN

It has long been feasible to preserve semen by freezing. Only a few years ago, when a young male university student wanted to raise some pocket money he could produce a semen specimen and sell it to a physician who had a patient requiring artificial insemination by donor (AID). The physician would ask the student a few questions regarding his general health, the possibility of recent exposure to venereal diseases, and the likelihood of his carrying hereditary traits that might jeopardize any offspring. The student might then be examined, and would undergo a few basic blood tests and a simple semen analysis. The physician probably would also document a few of the donor's physical characteristics, such as height, complexion, and eye and hair color.

As we mentioned earlier, the use of fresh semen is no longer an acceptable option. Sperm banks still obtain semen specimens from students as well as other donors, but now they compile far more extensive background information on the donor, including the test for exposure to AIDS and hepatitis. A masturbation specimen of semen is then frozen and stored for up to six months, at which time the AIDS antibody blood test is repeated. Only if the second test is negative can the sperm bank then safely release the specimen for use. The reason for this delay is that it can take several months for an individual infected with the AIDS virus to develop detectable amounts of antibodies in the blood. For these reasons, we urge couples to deal only with sperm cryopreservation banks.

Because it is no longer feasible from a medico-legal, moral, and ethical viewpoint to safely use fresh donor semen, frozen semen is today the only option for all forms of donor insemination, regardless of whether intravaginal or intrauterine insemination, IVF, or GIFT is to be performed. A rare exception to this rule might be justified when a woman insists upon selecting her own donor. In such cases, both the patient and the donor should be fully informed of all the potential risks involved when fresh semen is used. Like the semen donor discussed above, the donor designated by the aspiring mother should also undergo routine testing and evaluation, including tests for AIDS and hepatitis B. Thereupon, the legal agreements necessary to protect all parties from frivolous litigation should be consummated before donor insemination is performed.

The use of frozen semen impacts fertility treatment in several ways. First, freezing and transporting it over long distances in liquid nitrogen canisters is far more costly than using a fresh specimen. Second, although each specimen contains millions of sperm, a significant number die or lose their vitality and motility during the freezing and thawing process. However, even if half of the sperm in a semen specimen fail to survive cryopreservation, pregnancy can still occur because fertilization of an egg requires only one healthy sperm out of the millions contained within a specimen.

The pregnancy rate when thawed (previously cryopreserved) sperm are inseminated intravaginally does appear to be significantly lower than when fresh sperm are used. However, if intrauterine insemination is performed this raises the pregnancy rate to about 20% per cycle when conducted around the time of ovulation in natural cycles. These statistics are consistently reported across the board. Clearly, it is the way to go.

EMBRYO FREEZING

The dramatic advances in the technology of freezing and storing human embryos for future use have exciting implications for IVF. In the past, most IVF laboratories performed embryo cryopreservation only in selected cases where it was deemed that too many embryos had resulted from the IVF process than could safely be transferred to the woman's uterus. These leftover embryos, having undergone several stages of cleavage, were frozen once they comprised four to six blastomeres (cells). If the woman failed to conceive or if she conceived and subsequently wanted to attempt another conception, the embryos would be thawed at a selected time of a subsequent menstrual cycle, usually four or five days after the presumed time of ovulation, and would be transferred to the woman's uterus. This approach resulted in less than a 5% birthrate per embryo transfer.

The downside of freezing zygotes (fertilized eggs that have not yet undergone cleavage) or embryos is that about 20 to 30% of all cryopreserved reproductive cells, whether zygotes, blastomeres, or embryos, may be destroyed during the freezing and thawing process.

Recent evidence strongly suggests that there may be significant advantages to freezing zygotes or embryos at the earliest possible stage of cleavage. These zygotes and embryos would then be thawed at the appropriate time, cultured in a laboratory dish in order to ascertain their potential for continued cleavage, and would then be transferred to the woman's uterus during a natural cycle or following estrogen and progesterone hormonal replacement.

The process of embryo freezing requires the use of cryoprotectants that protect reproductive cells from destruction during the freezing process. Glycerol and dimethylsulfoxide (DMSO) are two cryoprotectants that were commonly used in the past. Currently, most programs use propanediol as a cryoprotectant. It appears to significantly reduce the reproductive cell attrition rate and may thereby significantly improve the chances of zygotes.

These advances in embryo freezing research offer great hope for the future. When this book was being written, the IVF birthrate at Pacific Fertility Medical Center in San Francisco was about 33% following the transfer of four or more thawed embryos into a woman's uterus. This process could potentially provide a couple with several opportunities to conceive following the performance of a single egg-retrieval procedure.

The time may well arrive when a woman, well stimulated with fertility agents so as to develop the maximum number of healthy follicles, would undergo a single ultrasound-directed egg retrieval procedure under local or light general anesthesia and the eggs would be fertilized in the laboratory. A few embryos might be transferred into the woman's uterus two or three days later, while the remaining zygotes or early embryos would be frozen, banked, and subsequently thawed in preparation for embryo transfer during one or more subsequent menstrual cycles, if necessary.

In our program we use thawed embryos only following hormonal treatment to build the uterine lining. As with preparing the uterus for IVF surrogacy and ovum donation, we once again favor the use of estradiol valerate and progesterone.

We believe that cryopreservation technology will continue to advance rapidly and will contribute significantly to the treatment of infertility in general and to successful IVF in particular.

EGG FREEZING

A few normal births have been reported following the transfer into a woman's uterus of an embryo or embryos derived from IVF of a previously frozen and then thawed human egg. It is, however, technically far more difficult to cryopreserve human eggs than embryos or sperm. Egg freezing technology is still in its infancy. There is currently a far greater attrition rate during the freezing and thawing of human eggs than is the case for sperm or even human embryos. This is particularly significant because a woman capable of producing only a limited number of eggs per cycle can ill afford to lose most of them during cryopreservation. The process exacts a lesser toll with sperm because a vast number of sperm survive even though many die during freezing. In addition, eggs are usually subjected to IVF during the same cycle of treatment in which they are retrieved, and it is unlikely that more than three or four eggs from any one couple would be left over and available for freezing after egg retrieval.

One of the reasons eggs are so much more sensitive to cryopreservation than sperm is believed to be the strikingly different composition of the two gametes. Sperm, the smallest cells in the body, comprise a large head containing the genetic material, and a tail. The egg, the largest human cell, resembles a chicken egg. In both the human and the chicken egg, the egg white is almost all ooplasm. The ooplasm contains the tiny microorganelles that nourish the fertilized egg and probably even the sperm once it has fertilized the egg.

Eggs are particularly sensitive to cryopreservation because freezing and thawing can produce two effects:

1. The fluid contained within the microorganelles in the ooplasm might expand during freezing, rupturing the microorganelles' membrane walls and thereby disrupting the metabolic processes within the egg. In contrast, the nuclear material that comprises most of each sperm appears to be relatively resistant to such damage.

2. The spindles from which chromosomes hang and that are indispensable to the exchange of genetic material during fertilization can be damaged when eggs are frozen and thawed. Spindle breakage could therefore potentially lead to an abnormal arrangement of chromosomes following fertilization.

However, it is important to keep in mind that nature is wisely selective. A defective egg almost certainly will not fertilize. If the embryo is defective, in the vast majority of cases it will not implant into the wall of the uterus, and the woman will never know that she had been pregnant. If the conceptus (the developing implanted embryo and/or early fetus) is damaged following implantation prior to the sixth to eighth week of pregnancy, the pregnancy will almost invariably abort. However, if an older conceptus or fetus is damaged, it is more likely that a birth defect might result. Nature makes a gallant attempt to maintain the integrity of the species by trying to ensure that defective gametes are incapable of fertilization, defective embryos do not implant, and imperfect conceptuses miscarry in the early stages of pregnancy.

Accordingly, an egg with defective ooplasm is highly unlikely to cause a problem, but the fact that spindle breakages have been observed following the thawing and fertilization of eggs has created some ambivalence on the part of many IVF specialists in regard to applying this technology in humans.

FROZEN EGGS VS. FROZEN EMBRYOS

The present survival rate for frozen embryos following thawing, based solely upon the observation that more than half of their cells appear to have weathered cryopreservation, is slightly greater than 70%. In contrast, the current chances are now well below 20% that an egg will survive the freezing process, with subsequent evidence of survival being apparently healthy cleavage. The recent introduction of newer methods promises to push this percentage up considerably in the foreseeable future. Although such numbers would seem to favor embryo freezing, it is possible that egg freezing, thawing, fertilization, and embryo transfer will ultimately prove to be far preferable to the use of embryos and could ultimately become standard practice in the IVF setting.

There are several theoretical advantages to cryopreserving eggs rather than embryos. First, an egg is a known quantity. It is likely to be healthy if it fertilizes and undergoes subsequent cleavage when thawed. In contrast, because the embryo is often transferred to the uterus immediately after thawing without undergoing further cleavage, one does not usually have the opportunity to observe whether it is indeed healthy at the time of transfer into the uterus. Because the embryo is further along the chain of evolution than an egg, the potential that an embryo damaged through freezing and thawing would produce an abnormal offspring is greater than that which could be anticipated from an embryo derived from a previously thawed egg.

Second, egg freezing skirts the ethical dilemma as to whether life in its earliest form is being manipulated. Eggs, like sperm, are considered to be cells that do not have life potential on their own. However, some people believe that an embryo represents the earliest form of life and that the 20 to 30% attrition rate of frozen-thawed embryos represents a form of abortion. However, the same argument cannot be applied to the freezing and thawing of eggs.

If one were to argue that it is unethical to freeze an egg because its chances of survival and subsequent fertilization are questionable, then it should likewise be unethical to freeze semen because many of the sperm also die during cryopreservation. In this context it might also be argued that the practice of vaginal intercourse is wasteful because only one or two, and rarely three or four, sperm might be capable of fertilizing eggs and producing offspring—and the remaining sperm would die. Thus, most people who have a religious or moral aversion to embryo freezing would be unlikely to have the same objection to the freezing of eggs.

Third, eggs are easier than embryos to obtain. Theoretically, women undergoing certain kinds of unrelated abdominal surgery might be willing to donate eggs. They could first be stimulated with fertility drugs and their eggs retrieved during surgery. These women could be reimbursed for donating their eggs to an egg bank or a waiting recipient, and the payment could help offset the cost of their surgery. Candidates might include women who undergo laparoscopy to have their fallopian tubes surgically occluded for the purpose of sterilization or women having hysterectomies for benign pelvic disease.

However, the largest source of eggs for ovum donation would be donors recruited through licensed agencies. This is already commonplace for the purpose of performing IVF with ovum donation. In addition, when more eggs are retrieved from a woman undergoing egg retrieval with IVF than are needed to optimize the likelihood that four to six embryos will be fertilized, the IVF couple might choose to donate or sell the surplus eggs to an egg bank or to other infertile women. We reject the sexist argument that it is immoral for women to sell their eggs while it is acceptable for men to sell their sperm.

Fourth, excess frozen eggs could be used to better diagnose male infertility where the cause could not be otherwise diagnosed. At the moment, hamster eggs and the hemi-zona test are used for diagnosis, but both have their flaws (see Chapter 9). Ultimately, as we have stressed repeatedly throughout this book, a couple's fertilization potential can only be assessed through examining the ability of the sperm to fertilize healthy eggs.

It is possible that the popularity of embryo freezing will ultimately be upstaged by egg freezing, but this is some time in the future. However, embryo freezing will always have a place in the IVF setting because of the likelihood that when large numbers of eggs are fertilized in the laboratory, the couple will be left with more embryos than they or the IVF team would be willing to transfer into the woman's uterus. These excess embryos would either have to be allowed to die spontaneously or be frozen, stored, and kept available for a subsequent chance at conceiving should the initial IVF cycle be unsuccessful.

Nevertheless, we believe that, with few exceptions, egg cryopreservation would be the better option in the future if it can be safely and successfully performed. As the technology continues to develop and be refined, we expect that egg cryopreservation will provide a variety of benefits to infertile couples.

MORE MORAL AND ETHICAL DILEMMAS: SHOULD NEW FERTILITY TECHNOLOGY BE REGULATED?

In addition to the questions raised by the cryopreservation issues, this new technology raises a host of other moral and ethical issues that have yet to be resolved—and probably never will be answered to everyone's satisfaction. The basic question is: To what extent should technology be allowed to alter the normal course of nature? In other words, where does it all end?

The following three examples illustrate the kinds of moral and ethical questions the laboratory director of a major IVF program encounters:

A 28-year-old female medical student asked:

Is it possible for you to freeze two or three stimulated cycles of my eggs now? I'll be over 35 by the time I get out of medical school, and I'd like to begin my family about age 40—but with age-28 eggs.

When a graduate student received widespread publicity after the birth of identical twin calves from a cow embryo he had split, a couple inquired of the laboratory director:

Would you please split one of our embryos so we can have identical twins?

And a terminally ill man asked:

My wife has agreed to bear me a large family after I'm gone. Would you freeze several samples of my sperm and artificially inseminate her over the years so she could have my family?

The medical director and other key staffers of an IVF clinic should fully expect to be confronted with many such requests in the future. Although addressing these dilemmas is not within the scope of this book, the examples mentioned throughout this chapter illustrate but a few of the moral and ethical issues that arise with IVF and related technologies.

Artificially Produced Embryos

Another approach that is theoretically possible would be injecting one or more blastomeres into a zona whose contents had previously been removed or enveloping the blastomeres with an artificial zona and then transferring the artificially produced embryos into a woman's uterus. It is also possible that human embryos potentially could be nurtured in the uterus of another species.

Pre-implantation Genetics (Genetic Engineering)

Approximately one in 500 babies born annually in the United States is afflicted with a sex-linked disorder (where only male infants get the disease), and one in 300 has a serious nonsex chromosome-linked abnormality. This means that almost 20,000 babies with chromosomal abnormalities are born annually in this country. In addition, it is estimated that one out of every 50 newborns has an undefinable major genetic abnormality. In other words, roughly 700,000 newborn babies are afflicted with an abnormality in some form each year in the United States.

We wonder how many couples whose first child was born with a major abnormality have opted to not have any more children because they don't want to risk bearing a second child with the same or another abnormality. And what of the couples who, knowing that they carry defective genes, opt to not have any children lest they pass the defect on to subsequent generations?

The emergence of *pre-implantation genetics* opens the door to some exciting possibilities for such couples. Although the science of genetic diagnosis and engineering is still in its infancy, we can say that the advent of IVF has made pre-implantation genetics possible because physicians now have genetic access to the human embryo.

It is already technically feasible to perform a biopsy on an embryo and remove a number of cells for genetic evaluation without reducing the subsequent likelihood of that embryo implanting once transferred to the woman's uterus. This means that we now have the opportunity through selectively transferring only healthy embryos to the woman's uterus to ultimately eradicate certain lethal diseases that have an identifiable genetic origin.

Ultimately, access to gene mapping, lists of all human genes and their whereabouts, will allow the diagnosis of many potentially lethal conditions prior to transferring embryos to the uterus. In selected high-risk cases, withholding the transfer of defective embryos to the uterus would reduce,the potential for genetic disease. Unfortunately, large-scale clinical access to pre-implantation testing is not likely to become the norm until early in the 21st century.

We are on the verge of nothing less than a biomedical revolution. There is no question that the diagnosis and treatment of gene-based disease will be the major medical challenge well into the 21st century. However, unlike many of the advances in reproductive technology that have suddenly exploded on the scene, the relatively slow development of pre-implantation genetics has allowed physicians, ethicists, lawyers, clergy, and citizens in general to discuss the implications in advance of the full implementation of the new technology.

ASSISTED FERTILIZATION (MICROMANIPULATION)

More than half of the infertility in the United States can be traced to a male factor. And in cases where male infertility is not amenable to correction using simple medical or surgical means, IVF offers the only rational form of treatment. Traditionally, the use of IVF in such cases has been associated with much lower success rates as compared to IVF performed for other causes. The main reason for this discrepancy is that spermatozoa obtained from infertile males are less likely to successfully fertilize eggs, regardless of laboratory techniques used to enhance their fertilization potential.

It is hardly surprising, therefore, that research in the '90s has been focused upon developing methods for promoting successful IVF in cases of severe male infertility. These procedures, which are collectively referred to as assisted fertilization or micromanipulation, require access to highly sophisticated technical expertise and equipment.

The three most widely used methods of such procedures involve: (1) creating a portal of entry for the sperm through the zona pellucida of the egg by dissecting away a portion of the zona (*partial zona dissection—PZD*); (2) injecting a number of sperm immediately below the zona but not into the cytoplasm of the egg (*subzonal insertion—SUZI*); and, more recently, (3) directly injecting a single sperm into the egg's cytoplasm (*introcytoplasmic sperm injection—ICSI*).

Partial Zona Dissection (PZD)

Here, the embryologist makes a cut in the zona of the egg, which is then placed in a sperm-rich environment. This creates a channel through which a relatively poorly motile sperm is able to enter the egg and achieve fertilization. PZD was the first attempt at achieving assisted fertilization and provided the first real insight into how we might improve fertilization potential in cases of male infertility. Fertilization rates using this technique range from 15 to 25%.

Subzonal Insertion (SUZI)

The introduction of PZD was soon followed by an attempt to inject sperm directly into the egg in cases where sperm motility was considered to be so poor that it was unlikely that fertilization could take place at all. With SUZI, a number of sperm are injected through the egg's zona pellucida into the perivitelline space surrounding the egg's substance. While fertilization rates using this technique are certainly higher than those which can be achieved through PZD, pregnancy rates remain relatively disappointing.

The significant drawback of SUZI is that it yields a relatively high incidence of polyspermia (in which more than one sperm fertilizes the egg, which in turns divides a number of times and then dies). In the conventional IVF setting, the incidence of polyspermia is about 3%, while with SUZI the incidence is several times greater.

Although SUZI produces pregnancies and is still widely used, it will undoubtedly be eclipsed by intracytoplasmic sperm injection (ICSI) in IVF settings that have access to the technical expertise and equipment necessary to perform this procedure. (See following section.)

Intracytoplasmic Sperm Injection (ICSI)

In this exciting and innovative procedure, a single sperm is captured by the tail using a very thin, delicate pipette and is injected directly into the cytoplasm of the egg. ICSI produces fertilization rates that approach 80% and birthrates that can equal or may exceed the success rates achieved through conventional IVF performed in the absence of male infertility. As such, this procedure has and will continue to impact profoundly on the entire field of in vitro fertilization.

The quality of the embryos achieved through ICSI appears to be good, and although an insufficient period of time has elapsed to enable full evaluation of the offspring that result from ICSI-induced pregnancies, there currently does not appear to be an increased risk of birth defects using this procedure. Presently, ICSI offers hope in the most severe cases of sperm dysfunction, even where sperm motility (ability to move about) is extremely poor or even non-existent. And in cases where the sperm count is less than one million per ml, ICSI consistently produces pregnancies. It is now even possible using ICSI to achieve pregnancies from biopsies taken from the testicles of men who for a variety of reasons do not ejaculate any sperm at all (azoospermia).

It is the authors' opinion that given the high fertilization and pregnancy rates associated with ICSI, eventually it will completely replace conventional methods of achieving IVF, even in cases where no male infertility exists. (See "Where Do We Go From Here?" in Chapter 15 for further discussion of ICSI and its implications for in vitro fertilization in the future.)

ASSISTED HATCHING

The zona pellucida has a complex structure that envelopes the embryo. In nature, about two days after an embryo reaches the uterus, the zona opens and out burst all the cells, which then try to burrow into the endometrium.

This is known as hatching. Assisted hatching is an IVF technique in which the zona is mechanically or chemically treated prior to embryo transfer in order to weaken the wall of the embryo and thus improve the likelihood of successful hatching.

Assisted hatching is most commonly used in women who have thickened zonas, which may occur more commonly with advancing age and in association with certain causes of infertility.

Zona Drilling

Embryologists first performed zona drilling by using an acidified solution to create a small hole in the zona that penetrates its entire thickness. Now an acid Tyrodes solution is often used to digest a small portion of the zona down to its innermost layer.

Protease Digestion (PROD)

In our setting we've developed a new method of assisted hatching using an enzyme to soften and thin the zona. The enzyme is called protease XIV, and the procedure is known as protease digestion (PROD). PROD gently softens the exterior of the zona in its full circumference in a manner analogous to natural hatching. We believe this approach is less traumatic to the embryo than either of the zona drilling procedures.

INTRAVAGINAL CULTURE FOR FERTILIZATION

In this recently developed variant of natural-cycle IVF (see Chapter 9), eggs and sperm are combined with culture medium in a small capsule, which is firmly packed into the warm, most environment of the woman's vagina. The woman is encouraged to be mobile and active for two days, whereupon the capsule is removed, and the fertilized eggs are processed and subsequently transferred into her uterus. The success rate of the fertilization is significantly less in this setting, and one is certainly limited to a small number of embryos that can be fertilized per attempt.

The success rate is no greater than that reported for natural-cycle IVF (10 to 15% pregnancy rate per cycle with an anticipated birthrate of 8 to 10% per cycle). Although touted by some as a cost-saver, this approach still requires embryologists, people to fertilize and identify the eggs, incubators to prepare the sperm, and so on. All it actually saves is a bit of laboratory space.

Nevertheless, this new culture method could play an important role in developing, emerging societies where access to high technology and the resources for developing major laboratory settings are lacking.

It also addresses some of the objections raised by religious groups that demand *in vivo* fertilization (fertilization in the vagina) even though it is in an artificial medium contained within the capsule. This approach may satisfy those groups that criticize the performance of IVF because it involves fertilization outside of the body.

SOME ETHICAL CONSIDERATIONS

In the early 1980s, a plane crash cut short the lives of an infertile couple, leaving two frozen embryos orphaned. The ethical and legal questions that arose seemed endless. Did the embryos have a right to be born? If so, did they inherit the couple's estate? Who would have decision-making powers over the fate of the embryos?

The answers to such questions must currently be formulated in the absence of clear ethical and legal guidelines to direct the use of IVF and related procedures. Because of this, leading professional organizations, such as the Ethics Advisory Board of the U.S. Department of Health, Education, and Welfare; the American Society for Reproductive Medicine (ASRM); the American College of Obstetricians and Gynecologists; and the Judicial Council of the American Medical Association (Ethics Advisory Board) are attempting to establish guidelines that would clarify the responsibilities of participants and set professional standards in the United States. All of the objectives dealt with by the various committees in this country and abroad are subject to a wide variety of interpretations. Furthermore, depending on demographics, geography, and the preponderance of different religious persuasions, ethical guidelines have to be modified to comply with acceptable standards within a particular community.

When one addresses the issues of morality and ethics in any particular IVF setting, it is imperative to examine not only the morality and ethics pertaining to sophisticated "Orwellian" developments in the field, such as embryo and gamete freezing, but also the "morality" of the entire technology. An example of the kind of question one could ask in this regard is, Is the treatment of infertility itself justifiable? or, Where does one draw the line in implementing the various reproductive technologies?

In the absence of legally enforceable ethical standards of practice in the United States, each IVF program has virtually free rein with regard to the manner in which standards of ethics and morality will be applied. Largely for that reason, we have established an Ethics Advisory Board in our clinics to review our ethical guidelines, to monitor our practical application of those guidelines, and to advise us in situations where difficult ethical decisions must be made.

The number of moral and ethical questions will only increase as new applications of artificial reproductive technology are introduced. Therefore, in order to answer those questions plus the ones that already confront us, we look forward to a day when clear ethical and legal guidelines directing the use of IVF and related procedures are available to everyone.

CHAPTER

15

PUTTING THE IVF
HOUSE IN ORDER

This book is based on the premise that IVF consumers (infertile couples and referring physicians) are at a great financial and informational disadvantage, and that the situation is not likely to change in the near future. If it is to improve at all, everyone who has an interest in IVF in the United States—physicians and others involved in IVF programs, insurance companies, fertility support groups, legislators, and IVF consumers—must work together in a concerted effort to help get the IVF house in order.

Many experts agree that there is a great need for the IVF community to deal with the consumer more openly. As far back as 1986, Dr. Gary Hodgen made the following statement at a medical conference:

I really believe that the public trust is the single greatest factor that has allowed the miracles of medicine to evolve in the twentieth century. . . . The public has allowed us a great deal of latitude to decide where we are going to go and how we are going to get there. I don't believe we have in all cases returned that respect with an equal degree of explanation and understanding, speaking to the fears and concerns of the public in general. Certainly we are not of a single mind among ourselves as to the appropriate course or end-point in decision-making with regard to the ethics of in vitro fertilization therapy and research.

We believe this statement to be as true today as it was then.

How can the IVF medical community respond to that public trust? We believe the first step would be to make IVF more accessible to all consumers. IVF programs could work toward this goal by (1) cooperatively standardizing procedures so consumers can expect about the same success rates wherever they go, (2) willingly providing reliable and understandable data to consumers, and (3) working with insurance companies and legislators to make the process affordable.

A proactive approach toward compiling and disseminating IVF information on the part of the medical community would go a long way toward educating members of the media, who all too often misunderstand and consequently misrepresent what IVF is all about. We are faced with too many contentious newspaper editorials and oversimplified TV news reports that paint an inaccurate and sometimes alarming picture about the success of IVF. Reversing the harmful trend of bad press by being openly accountable is one giant step that could be undertaken immediately.

CONSUMERS HAVE THE *RIGHT* TO EXPECT MINIMUM STANDARDS IN ALL PROGRAMS

Standards must be established for IVF programs in the United States, and consumers must have easy access to understandable data about success rates. In almost all other medical disciplines, consumers can safely assume that the physician who is going to perform a certain procedure has, or has access to, the required expertise. This should be true with IVF programs as well.

Consumers deserve to have similar outcomes from every IVF program in the United States. It is unacceptable that certain programs can promise a birthrate in excess of 30% per treatment cycle while others report less than half this success rate—or have no track record at all on which to base any statistical analysis. Is it right that a couple should pay such a huge amount when they don't know what their chances are?

One way in which IVF programs can meet minimum standards is by learning from and replicating proven programs. The general factors that contribute to a successful IVF program can be viewed as a triangle, with each side of the triangle representing a crucial ingredient: (1) technical expertise, (2) proven clinical and laboratory protocols and techniques, and (3) rigid quality assurance. The people who make an IVF program effective constitute the glue that holds the sides of the triangle together: commitment, teamwork, and determination are essential ingredients for the successful IVF program.

The structural integrity of this triangle might be compared to the interdependence between a lock and key. Once established (i.e., the IVF program functions effectively and the key opens the lock time after time), the winning combination should not be weakened through needless, ill-conceived tinkering. In the IVF program, as in the lock-and-key example, there is zero tolerance for deviation from a successful relationship. Just as it would be silly to file away at a key that fits a lock perfectly, it is also shortsighted to refuse to take advantage of state-of-the-art technical expertise, proven protocols and techniques, and unwavering quality assurance in the IVF setting. The technology exists. Lock-and-key IVF procedures can be replicated at many sites, enabling settings that adopt them to standardize their programs and consequently their success rates.

The difference between a poor and an excellent IVF program may have nothing to do with availability of expertise, equipment, or technical know-how. It simply may be the way in which the components are put together. In the case of a poor program, all the components may be in place, including technical expertise, but the program may be so poorly administered that there is no uniformity of outcome. An IVF program that doesn't have any set protocols and procedures might, with luck, come up with some good outcomes over a period of time—some good results in simple cases and worse results in difficult cases, with no consistency in success rates. A consistently successful IVF program will, depending on individual requirements determined beforehand, arrange the same components differently, but the components will be the same. A consistently successful IVF program can repeat the same recipe for success over and over and still adopt a reliable format for cases whose special circumstances require a special approach. The only time change would be called for would be in the introduction of new technology that improves the process.

We strongly believe that there should be a way to guarantee that the components of successful IVF programs can be replicated everywhere. Consumers should be able to have confidence that no matter where they live, they will have access to a program offering the same success rates as all others. The only way to accomplish this is by ensuring that all IVF programs use practiced and proven formulas for success, and that all results are validated and available to the public. Adoption of such techniques would be a giant step forward.

CONSUMERS HAVE THE *RIGHT* TO AFFORDABLE IVF

The high cost of IVF confronts consumer, physician, and insurance company with this chicken-and-egg situation: IVF is expensive because it is a relatively new, high-tech procedure; however, a greater volume of IVF consumers could lower both fixed and variable costs; but, few customers can afford IVF because most insurance companies will not cover it; and, insurance companies are reluctant to reimburse for IVF because the success rates vary so widely and there is no accountability; therefore, IVF continues to be prohibitively expensive because . . . and the cycle continues.

The reluctance of most members of the insurance industry to cover IVF should be viewed from their perspective: they are unwilling to accept the current statistics on clinic success rates. There are many reliability problems with current statistics. There is no universal method for measuring success. Only with accurate data can insurers calculate their risk and decide on a fair premium for this kind of coverage. Then, and only then, will IVF be covered by insurance companies as, for example, tubal surgery is now.

WHAT CAN BE DONE TO REDUCE THE COST OF IVF?

We believe that because fewer than 35,000 procedures (out of the pool of more than one million potential IVF couples) are currently performed yearly in the United States, most of the 250 or so programs in this country are grossly underutilized. The number of procedures performed barely scratches the surface of the demand. One-half of all the IVF procedures in the United States are probably performed in fewer than 30 programs, with the remainder being divided among the remaining clinics. Since some larger programs are doing more than 500 procedures a year, that means that others are performing far fewer than 100. Yet no one can gain optimal expertise doing so few procedures per year. It is impossible even to develop statistics, let alone confidently report them, when they are based on such small samples.

Most important, consumers must be attracted to IVF because of its reliability and quality. But merely interesting more consumers in the concept of IVF is not enough—the procedure must be made affordable, which brings us back to the issue of medical insurance coverage.

INSURANCE COVERAGE IS THE KEY TO IVF AFFORDABILITY

During an appearance on the "Oprah Winfrey Show" in 1993, Dr. Sher made the following observation about the double standard that exists today with regard to insurance reimbursement for certain fertility treatments in the United States:

> *When most insurance companies reimburse for procedures such as penile implants done in cases of male impotence and yet refuse to cover infertility, it makes one wonder how many directors and CEOs of these companies are older men who view male impotence as a life-endangering condition and the desire of a woman to have a baby as a vanity.*

This double standard also applies to reimbursement for tubal surgery, although IVF in the proper setting invariably offers a far greater chance of pregnancy than tubal surgery. Until insurance companies change their outlook, this double standard will be perpetuated.

What will it take to obtain universal insurance coverage for IVF? As we have said all along, before insurance companies are likely to cooperate, IVF programs must openly account for their success rates. All United States IVF programs should submit their statistics on quality of service for review by an impartial accrediting agency.

While we are not making a bid for mandatory regulation, we do advocate independent accreditation of IVF programs. There are strong incentives for IVF programs to participate in such accreditation, including the increased likelihood for insurance reimbursement, which then would lead to greater patient volume. In turn, higher patient volume would reduce variable costs and lower overhead by allocating fixed costs over a larger patient base, thus lowering overall costs to couples.

The second approach to obtaining insurance coverage for IVF is for government at the state or federal level to require that insurance companies cover the procedure. As of 1994, at least 10 states in the United States had introduced and many had passed such legislation. However, one must question whether this is the best approach, because legislation will not necessarily promote accountability. It merely mandates funding.

We believe that accountability and legislation should go hand in hand. Neither approach would be entirely successful alone. But it will not be easy to accomplish these changes. Convincing insurance companies that it is in their own best interests to fund IVF performed by accredited programs will be a long, slow process. And until there is more accountability by individual IVF programs, the insurance companies forced to fund procedures with widely varying outcomes might be expected to lobby for repeal of mandatory reimbursement laws.

Therefore, the quickest and most effective way to make IVF affordable for the majority of the American public is through a combination of approaches: increased accountability by the medical profession and the insurance industry, legislative mandate, and consumer pressure.

ACCREDITATION OF IVF PROGRAMS WOULD BENEFIT INFERTILE COUPLES AND THE MEDICAL PROFESSION

Central to the entire issue of inequitable insurance reimbursement policies for IVF and related advanced reproductive technologies is the concept of accountability—by the medical profession, in regard to reported success rates, and by the insurance industry, with respect to funding procedures performed by IVF and related programs that meet acceptable standards of outcome based on average national statistics.

We believe that the only way we can bolster faltering consumer confidence in IVF is by means of accountability as administered and monitored by an accrediting body. Such an accrediting body might seek advice from industries that market fertility and fertility-related medications, diagnostics, and equipment. The insurance industry, IVF and related programs, infertility-oriented consumer groups, and perhaps even various levels of government are examples of groups that would all benefit from accountability. This body might work closely with ASRM (American Society for Reproductive Medicine), SART (the Society of Assisted Reproductive Technology), and the American College of Obstetricians and Gynecologists, all of which have already addressed the issue of operational standards as they should pertain to IVF and related technologies.

Although acceptable standards of outcome are difficult to quantify, the IVF success rates published annually in the journal *Fertility and Sterility* by SART provide at least a starting point. The 1995 SART report presented the 19% nationwide average birthrate per egg retrieval for women under the age of 40 as compared with a 6% birthrate for women over 40. Accordingly, these birthrates might be considered acceptable standards for measuring success with IVF.

While the clinical pregnancy rate per egg retrieval is an indication of an IVF program's proficiency, it is the birthrate-per-egg-retrieval statistic that is most meaningful to consumers because it best predicts the outcome they might realistically expect (the actual "baby take-home rate" per egg recovery).

Under the system we propose, IVF programs might submit themselves to an ongoing process of peer review. Participating programs would register each prospective patient with the accrediting body prior to initiation of treatment. A patient code number could ensure confidentiality, and registration of the patient with the society would guarantee proper data interpretation.

We believe in this proposal so strongly that in 1993 we voluntarily submitted our programs in Pacific Fertility Medical Center to an accreditation audit undertaken by a nationwide auditing firm.

In accordance with the rules of the auditing process, our program submitted the demographic and clinical data pertaining to all of our candidates for IVF to the auditing firm prior to the day of egg retrieval. All of the data were held and the outcome was subsequently investigated by the auditing firm, which reported the validated results. In this way, we committed ourselves to being accountable for the success rates we quote.

These success rates were reported in two categories by age, one for women over the age of 40 and the other for women under 40. They were further subdivided on the basis of the absence or presence of male factor infertility, which at that time (prior to the advent of ICSI) was expected to lower the success rates.

Newer programs that initially submit themselves for accreditation would have 12 months to demonstrate that they are indeed able to meet "acceptable standards" of pregnancy and birthrates. Programs wishing to apply for accreditation after the first year could submit to a similar prospective evaluation or might alternatively elect to undergo a detailed retrospective audit according to the standards set forth by the accrediting body's peer-review committee. Instead of eliminating marginal IVF programs, which might occur as a result of government-mandated regulation, such an accrediting body would set an example and would even help struggling programs upgrade their standards and performance in the area of high-tech infertility treatments.

Each accredited program would undergo an annual peer review to become reaccredited. This would provide an ongoing assurance of proficiency to the consumer and to the referring doctor, and would also give each program important feedback regarding its own performance.

The attainment of accreditation by a program might prompt insurance companies to reimburse for IVF and related procedures performed in that particular setting. We anticipate that participation in such an accrediting process would snowball as IVF programs become convinced that accreditation would be in their own best interest for the sake of insurance reimbursement and to forestall mandatory regulation by federal or state governments.

In our opinion, the IVF registry sponsored by SART (see Introduction) has been an admirable but relatively inadequate effort to achieve quality assurance and set standards for IVF and related programs. In the past, the lack of an autonomous peer-review system for reported results perpetuated questions about the credibility of data submitted and did nothing to dispel suspicion by the medical profession, the insurance industry, and infertile couples regarding reported success rates. SART is currently in the process of revisiting the issue of establishing a prospective and verifiable audit of IVF success rates for all its member clinics.

The decision by SART to provide information on a clinic-specific basis since 1990 was certainly a step in the right direction. But their failure to undertake a prospective peer-reviewed audit prompted Oregon congressman Ron Wyden to introduce legislation to mandate such accountability. The Fertility Clinic Success Rate and Certification Act, HR 4773, was signed into law in late 1992. This bill mandated prospective audits and took effect in 1995. Although this important and much-needed legislation is far from perfect, it certainly represents an important first step in the right direction. The Lab Certification Act, which became law in November 1992, required monitoring IVF laboratories beginning in 1995.

Unfortunately, the federal government's intervention and SART's renewed willingness to revisit the issue of accountability comes too late to help the millions of infertile couples who couldn't afford the procedure in the past; but it may result in program accountability acceptable to insurers, and will ultimately make IVF far more affordable. We hope this will eliminate the aura of skepticism that currently threatens the existence and growth of this valuable infertility technology.

IVF CONSUMERS HAVE AN *OBLIGATION* TO GET INVOLVED

Ultimately, consumers can control the debate. They may have to band together to make their voices heard against the forces of the marketplace, but they can bring about change. Now is the time for IVF consumers to be outspoken. If they don't participate in the campaign to put the IVF house in order, they have only themselves to blame if progress comes slowly. One of the most promising lobbying avenues would be to join one of the infertility support groups, both to become more informed and to speak with a louder voice before the medical profession, legislative groups, and the insurance industry.

It is time for consumers to marshal their buying power to demand that these "big A's" in the field of high-tech infertility management are met:

1. **Accreditation** of IVF programs

2. **Accountability** by the medical profession with regard to providing validated statistics or a track record, and instilling rational expectations in infertile couples who seek their advice

3. **Availability and access** to the consumer of state-of-the-art standards of care

4. **Affordability**

WHERE DO WE GO FROM HERE?

After an initial shakeout period following accreditation, the United States may have fewer IVF programs, but they will be programs validated by peer review and offering a uniformly reliable success rate. Just because there are fewer programs, however, does not mean that access to IVF will be more restricted than it is now for consumers who do not live in metropolitan areas. On the contrary, mobile units could bring IVF and related procedures to the couple's own area, where they are familiar with the doctor and feel most comfortable.

For example, instead of 25 small programs in one geographical area, all of which have relatively high costs because they cannot benefit from economies of scale, consolidation and regionalization might provide better service to the entire area. One large, well-equipped center could serve outlying communities as well as the metropolitan area, reducing overhead costs while maintaining an optimal level of technology and research.

HOW WILL IVF BE PERFORMED IN THE FUTURE?

As we mentioned in the previous chapter, we believe that ICSI (intracytoplasmic sperm injection) is a phenomenal procedure. In our own setting, it plays an indispensable and ever-increasing role. The implications of ICSI promise to revolutionize many aspects of in vitro fertilization as we know it today. ICSI by its very nature is a high-tech, highly sophisticated procedure and is extremely costly to perform. The average IVF program affiliated with a physician's private practice can simply not afford to provide this service, at least not in the foreseeable future. However, ICSI has already become widely employed in large IVF centers that have access to high-tech personnel and equipment.

As stated in an earlier chapter, we are convinced that the ICSI technique will ultimately replace conventional methods of achieving in vitro fertilization even when male infertility is not a problem for the IVF couple.

Now we will examine some of the reasons why we believe that ICSI will revolutionize the performance of IVF.

ICSI Is Effective and Efficient

So many factors may inhibit fertilization (e.g., in addition to the ones we discussed earlier there are such considerations as subtle abnormalities affecting the surface of both the egg and the sperm), it would be easier and certainly more efficient to just sidestep these problems by direct fertilization via ICSI.

ICSI Enables Sperm of Various Qualities to Fertilize the Egg

As you know, it only takes a single sperm to fertilize an egg. With ICSI, even very poor quality sperm can be successful. They don't even have to be young. All that needs to happen in order for fertilization to occur is for the sperm DNA package to gain access to the egg's cytoplasm. Successful fertilization and pregnancies have been recorded in cases where ICSI has been performed using completely immotile sperm. And there's really no evidence to suggest that sperm looking and moving differently from other sperm are in any way more likely to have an abnormal DNA package.

The performance of ICSI *theoretically* removes all the handicaps to spontaneous fertilization and potentially creates a level playing field for achieving fertilization in all forms of infertility. What this means is that with ICSI virtually all men who produce sperm could potentially fertilize an egg because now it truly would take only one sperm to do so. We believe that in the future all IVF will be conducted with ICSI because of its obvious advantages. However, the high-tech requirements of ICSI are unlikely to be affordable to small IVF clinics in the near future. Centers offering ICSI will likely spring up regionally in areas that can financially support them. Optimally, smaller IVF programs would be able to work in affiliation with such institutions to bring access to ICSI into less-populated areas.

From the Petri Dish to the Test Tube

We expect that IVF laboratories will soon move from conventional methods of fertilization in a petri dish (see Chapter 2, "The Process of Fertilization") to culturing eggs in suspension media in an actual test tube. Cells or secretions of cells derived from the lining of the human fallopian tube or from other female reproductive organs will likely supplement or completely replace current embryo growth media, thereby improving the environment for fertilization and cell division. This is also bound to improve fertilization rates and embryo quality in the future.

The Big Question: Will Fertility Drugs Become Unnecessary?

We believe that we are fast coming to the end of the era when fertility drugs are indispensable to the removal of a sufficient number of eggs from a woman's ovaries. In the mid-1990s a group at Monash University in Melbourne, Australia, demonstrated the possibility of retrieving numerous healthy eggs from women who had *not* received any fertility drugs at all in advance of the egg retrieval. This procedure is known as *immature oocyte (egg) retrieval*. (To refresh your memory about egg growth and development, please see Chapter 2, "The First Half of the Cycle.")

Within 10 to 12 days of the onset of normal menstruation, these researchers performed an ultrasound needle aspiration of the very small follicles in the woman's ovaries and removed immature eggs. These eggs were subjected to a complex process of maturation in the laboratory and then fertilized using ICSI. The resulting embryos were deposited in the woman's uterus a few days later.

This exciting new procedure offers tremendous promise for the future. It allows potential access to a large number of healthy eggs without requiring the prior administration of fertility drugs. This is a very noble approach in view of the potential hazards and the exorbitant cost associated with the administration of fertility drugs.

Immature oocyte retrieval is also expected to become popular with egg donors, who would not have to submit to a regimen of fertility drugs prior to undergoing egg retrieval for the purpose of donating their eggs. There is little doubt the days of routinely using fertility drugs so as to ensure that a number of eggs can be harvested at any given egg-retrieval procedure are fast coming to an end.

THE BOTTOM LINE

Ultimately, society itself must determine whether technology should be allowed to run rampant or allowed to progress in a controlled manner. We recognize the widespread concern of the medical community that regulation of one aspect of medicine may lead to creeping regulation of the entire profession. Nevertheless, we believe it would be socially responsible to adopt, at the national level, directives or requirements that would control this developing technology for the public good.

The social responsibility that confronts practitioners of IVF and related technologies was underscored by Dr. Gary Hodgen at a 1985 IVF conference in Reno when he said

How can we in the area of in vitro fertilization do anything other than search and struggle together to find what this moral and ethical obligation is, define it, and attempt to refine it as we move forward with research results, technology, and new capabilities to help infertile couples?

There is little difference of opinion in this pluralistic society about the needs of people to have well children. The issue that's at risk is how we get there—implementation of research and clinical care.

The time has come to move from recommendations and guidelines, inconsistently applied, to strong directives that can be enforced. Loosely stated guidelines do not provide enough guidance, and they leave the field wide open to abuse. All too often, guidelines have been adopted because decision makers are afraid to say "This is what you *will* do" and thus settle for "This is what we recommend you might do *if you want to.*"

Decisions about the future directions of fertility technology cannot be left to one interest group. In our pluralistic society, varying viewpoints and backgrounds must be represented in order to make the consensus process work: consumers, physicians and other practitioners of IVF, the clergy, fertility support groups, lawyers, insurance carriers, ethics specialists, the media, and legislators must all work together to bring about the national adoption of comprehensive, enforceable directives to guide the implementation of research and clinical care in the field of infertility.

For the thousands of couples whose lives have been enriched by the gift of life through IVF and the other assisted reproductive technologies, for the many more infertile couples who have little hope of conceiving without the assistance of these procedures, and mindful of the sacred doctrine that obliges the medical profession to improve the human condition and alleviate suffering wherever possible, we challenge the medical/scientific communities and consumers alike to strive together to expand the technology, improve the quality, and promote the affordability and accessibility of IVF and related technologies.

GLOSSARY

acid tyrodes digestion A form of assisted hatching in which the embryo is introduced into a chemical solution that partially erodes the zona (egg covering) in order to promote hatching.

acrosome The protective structure around the head of the sperm. The acrosome contains enzymes that enable the sperm to penetrate the egg.

acrosome reaction The second stage of capacitation, when a sperm sheds its outer membrane to expose receptors that interact with the egg's zona pellucida to initiate fertilization.

adenomyosis A condition in which the endometrial glands grow into the uterine wall, creating a spongelike effect; can be associated with poor uterine linings. This condition is sometimes associated with heavy, painful periods and uterine enlargement.

adrenal glands Small structures located at the top of each kidney that produce a number of hormones indispensable to proper growth, development, and a wide variety of physiologic functions.

AF See assisted fertilization.

AID *See* artificial insemination by donor.

AIDS A sexually transmitted disease believed to be caused by one or a variety of viruses that are harbored in the nuclei of cells and attack the immune system. Infected individuals become highly susceptible to opportunistic infections; AIDS ultimately leads to death.

alloimmunity Immunity that develops against the proteins of another individual of the same species.

alphafeto protein A chemical in the blood and amniotic fluid that if found might point toward a neurologic fetal malformation.

American Fertility Society Former name of the American Society for Reproductive Medicine (ASRM).

American Society for Reproductive Medicine (ASRM) A professional society that primarily includes physicians but also includes laboratory personnel, psychologists, nurses, and other paramedical personnel interested in infertility. Formerly known as the American Fertility Society.

antibodies to sperm Substances in the man's or woman's blood and in reproductive secretions (semen, uterine and tubal secretions, and cervical mucus) that reduce fertility by causing sperm to stick together, coating their surface or killing them.

antilymphocyte antibodies (ALA) Antibodies formed to combat the male partner's lymphocytes and hence against the fetus's lymphocytes.

antiphospholipid antibodies (APA) Antibodies to some of the chemical substances that coat the root system of the placenta as it grows into the uterine wall. Women with high concentrations of these substances may have a higher incidence of miscarriages or may fail to conceive after repeated attempts.

anus Excretory opening of the intestinal tract.

ART See assisted reproductive technology.

artificial insemination by donor (AID) The most common form of insemination into the vagina or uterus; AID involves the use of donor semen or sperm in cases where the woman's partner is infertile or the woman chooses to conceive without having intercourse with the sperm provider.

ASRM See American Society for Reproductive Medicine.

assisted fertilization (AF) Methods for promoting successful IVF in cases of severe male infertility; these approaches require highly sophisticated technical expertise and equipment. Also known as micromanipulation.

assisted hatching A technique in which the zona pellucida (outer shell of the egg) is chemically or mechanically thinned prior to embryo transfer in order to improve the likelihood of subsequent hatching.

assisted reproductive technology (ART) Procedures involving retrieval of eggs, and the enhancement of eggs and sperm outside the body. It includes procedures such as gamete intrafallopian transfer (GIFT), in vitro fertilization (IVF), and zygote intrafallopian transfer/tubal embryo transfer (ZIFT/TET).

augmented laparoscopy A procedure in which eggs are retrieved from the woman's ovaries while diagnostic laparoscopy is being performed to evaluate the integrity of her pelvic organs. These eggs are subsequently fertilized in vitro, and the embryos are transferred into the woman's uterus two or three days later. This procedure affords a woman undergoing routine diagnostic laparoscopy a chance to determine the cause of her infertility and an opportunity to conceive by IVF at the same time.

autoantibodies Antibodies that are formed against the proteins of the individual's own body.

basal body temperature (BBT) chart A daily body temperature chart that provides a rough idea of when ovulation occurred. This is possible because body temperature rises when the corpus luteum produces progesterone (after ovulation) and drops at or just before the beginning of menstruation, when estrogen and progesterone levels fall (see also biphasic pattern of temperature on BBT chart).

BBT chart See basal body temperature chart.

Billings Method of contraception A method of predicting ovulation in which the woman examines the quality and quantity of her cervical mucus secretions. This method can be used to help the woman determine her most fertile period for the purpose of conceiving or for contraception.

biphasic pattern of temperature on BBT chart Charting pattern that occurs because the woman's temperature is likely to be ½° to 1° lower during the first phase of her menstrual cycle than during the second half, when the progesterone produced by the corpus luteum raises her temperature slightly (see also basal body temperature chart).

bladder The anatomical reservoir that receives urine produced by the kidneys.

blastocyst An advanced stage of embryo development during which a cavity develops within the young embryo.

blastomere Cell within the developing embryo. Each blastomere is capable of developing into an identical embryo until the embryo reaches about the 30-cell stage, after which the cells begin to differentiate into specific tissues.

blood-hormone test—LH When this test is performed several times daily around the presumed time of ovulation, the detection of a rapidly rising blood LH (luteinizing hormone) concentration can accurately determine the time of probable ovulation. This test, which requires blood to be drawn several times and is therefore painful, time-consuming, and expensive, has been virtually supplanted by serial urine LH testing (see also urine ovulation test).

blood-hormone test—progesterone Measuring of the concentration of progesterone in the woman's blood during the second half of the menstrual cycle about one week prior to anticipated menstruation; indicates whether or not she is likely to have ovulated because progesterone is usually produced only by the corpus luteum, which develops after ovulation.

capacitation The process by which sperm are prepared for fertilization as they pass through the woman's reproductive tract (in vivo capacitation); sperm may also be capacitated in the laboratory (in vitro capacitation).

cervical canal The connection between the outer cervical opening and the uterine cavity.

cervical mucus Mucus produced by glands in the cervical canal; it plays an important role in transporting sperm into the uterus and in initiating capacitation.

cervical mucus insufficiency A condition in which the ability of the cervical mucus to initiate the capacitation process is compromised through a deficiency in the amount of mucus produced, an abnormality in the physical-chemical components of the mucus, the presence of infection, an abnormal hormonal environment, or the secretion of antibodies to sperm in the mucus. Cervical mucus insufficiency is responsible for about 10 percent of all cases of infertility.

cervix Lowermost part of the uterus, which protrudes like a bottleneck into the upper vagina; the cervix opens into the uterus through the narrow cervical canal.

chemical pregnancy Biochemical evidence of a possible developing pregnancy based on a positive blood or urine pregnancy test; at this point, pregnancy is presumptive until confirmed by ultrasound (*see also* clinical pregnancy).

chlamydia Bacteria responsible for a sexually transmitted infection that may damage the fallopian tubes and/or the male reproductive ducts, thereby causing infertility.

chromosomes Structures in the nuclei of cells, such as the egg and sperm, on which the hereditary or genetic material is arrayed.

classic surrogacy The use of a third party to conceive and carry a baby to term. In this form of surrogacy, the baby would bear the genetic imprint of the surrogate and of the sperm provider.

cleavage The process of cell division.

climacteric The hormonal change that precedes the menopause by a number of years and is associated with a progressive loss of fertility, an increased incidence of abnormal or absent ovulation, hot flashes, irregular menstruation, a progressive rise in blood FSH levels, and mood changes. The climacteric usually represents an important stage in a woman's life.

clinical pregnancy A pregnancy that has been confirmed by ultrasonic examination or through pathologic assessment of a surgical specimen obtained either from a miscarriage or from an ectopic pregnancy. A clinical pregnancy should be distinguished from a chemical pregnancy, which through a positive blood pregnancy test merely suggests the possibility that a pregnancy has occurred.

clitoris The small structure at the junction of the labia minora in front of the vulva. The clitoris, which is analogous to the penis in the male, undergoes erection during erotic stimulation and plays an important role in orgasm.

clomiphene citrate A synthetic hormone that is used alone or in combination with other fertility drugs to induce the ovulation of more than one egg. When marketed in the United States, clomiphene citrate is also known as Clomid or Serophene.

COH *See* controlled ovarian hyperstimulation.

conception Creation of a zygote by the fertilization of an egg by a sperm.

conceptus A term used to describe the developing implanted embryo and/or early fetus.

controlled ovarian hyperstimulation (COH) In response to the administration of fertility drugs, the maturation of several follicles simultaneously, which results in the production of an exaggerated hormonal response.

corona radiata *See* cumulus granulosa.

corpus luteum A term for a follicle after an egg has been extruded. After ovulation, the follicle collapses, turns yellow, and is transformed biochemically and hormonally. The corpus luteum produces progesterone and estrogen, and has a life span of about 10 to 14 days, after which it dies unless a pregnancy occurs. If the woman becomes pregnant, the life span of the corpus luteum is prolonged for many weeks. A synonym for the corpus luteum is the "yellow body."

cryopreservation The process of freezing (in liquid nitrogen) and storing eggs, sperm, and embryos for future use.

cul-de-sac Area of the woman's abdominal cavity behind the lower part of the uterus.

cumulative birthrate The overall chance of a woman having one or more babies per egg retrieval or per embryo transfer following several attempts.

cumulative pregnancy rate The overall chance of a clinical pregnancy occurring per egg retrieval or per embryo transfer following several successive procedures.

cumulus granulosa The group of ovarian cells resembling a sunburst that surround the zona pellucida of the human egg; also called the corona radiata. These cells nurture the egg while in the fallopian tube.

de Miranda Institute A consumer protection agency for infertile couples.

DES (diethylstilbestrol) A drug previously taken by women during pregnancy that may cause infertility and/or pathologic conditions in the reproductive tracts of both male and female offspring.

diagnostic hysteroscopy A procedure usually performed under local or general anesthesia in the office setting or in the operating room. A thin telescope-like instrument is inserted via the vagina and cervix into the uterine cavity. Carbon dioxide gas or a liquid is injected to distend the cavity and allow direct visualization of its structure.

diagnostic IVF The performance of in vitro fertilization for the purpose of assessing the ability for fertilization to take place. It is an objective test of sperm/egg fertilization potential.

E$_2$ *See* estradiol.

ectopic pregnancy A pregnancy that occurs when the embryo implants in a location other than the uterus; the most likely site for such implantation is the fallopian tube (in which case the term ectopic pregnancy is used synonymously with tubal pregnancy). If undetected, an ectopic pregnancy may rupture and cause life-threatening internal bleeding. Ectopic pregnancies almost always require surgical intervention.

egg The female gamete, which develops in the overay; also known as an ovum or oocyte. An egg is the largest cell in the human body.

egg retrieval The retrieval of eggs from the ovarian follicles prior to ovulation; the eggs are sucked out of the follicles through a needle either during ultrasound guidance or laparoscopy.

ejaculation The emission of semen through the urethra and penis that follows erotic stimulation and accompanies male orgasm.

embryo The term for a fertilized egg from the time of initial cell division through the first six to eight weeks of gestation. Thereafter, the embryo begins to differentiate and take on a human organic form; at this point it is traditionally referred to as a fetus.

embryo adoption This occurs when a woman receives into her uterus an embryo to which neither she nor her partner has contributed a gamete.

embryo co-culturing The addition of cells derived from the growth of other tissue (from the lining of human or bovine fallopian tubes, or human follicular lining) to the culture medium in which the zygote is being nurtured in the laboratory. This is thought to enhance growth and promote the development of healthier embryos.

embryo transfer The process whereby embryos that have been grown in vitro are transferred into the uterus.

endometrial biopsy Surgical removal of a specimen of the endometrium, commonly performed to permit microscopic examination of the effect of estrogen and progesterone on the endometrium. When performed by an expert, it is usually possible to pinpoint almost to the day when ovulation is likely to have occurred.

endometriosis A condition in which the endometrium grows outside the uterus, causing scarring, pain, and heavy bleeding, and often damaging the fallopian tubes and ovaries in the process. Endometriosis is a common organic cause of infertility.

endometrium The lining of the uterus, which grows during the menstrual cycle under the influence of estrogen and progesterone. The endometrium grows in anticipation of nurturing an implanting embryo in the event of a pregnancy; it sloughs off in the form of menstruation if implantation does not occur.

epididymis Tubular reservoir that contains and transfers sperm to the vas deferens and subsequently through the urethra and penis at the time of ejaculation.

estradiol (E_2) A female hormone produced by ovarian follicles. The concentration of estrogen in the woman's blood is often measured to determine the degree of her response to controlled ovarian hyperstimulation with fertility drugs. In general, the higher the estradiol response, the more follicles are likely to be developing and, accordingly, the more eggs are likely to be retrieved.

estradiol valerate A preparation of natural estradiol taken orally or by injection.

estrogen A primary female sex hormone, produced by the ovaries, placenta, and adrenal glands.

exit interview An interview prior to the couple's release from an IVF program after the performance of embryo transfer, GIFT, artificial insemination, or related procedures. An exit interview prepares the couple for their return home and provides valuable feedback to the program.

extracorporeal fertilization Synonym for in vitro fertilization.

fallopian tubes Narrow 4-inch-long structures that lead from either side of the uterus to the ovaries.

falloposcope A telescope-like instrument that is introduced into the fallopian tubes for diagnostic purposes during falloposcopy.

falloposcopy A procedure performed at the time of laparoscopy or hysteroscopy, in which a thin telescope-like instrument is introduced into the fallopian tube to evaluate its condition.

fertility drugs Natural or synthetic hormones that are administered to a woman in order to stimulate her ovaries to produce as many mature eggs as possible, or to a man in an attempt to enhance sperm function or production.

fertilization The fusion of the sperm and egg to form a zygote (*see also* zygote, conception).

fetus Once the embryo differentiates and begins to take on identifiable humanlike organic form, it is termed a fetus; the fetal stage of development usually begins around the eighth week of pregnancy.

fibroid tumor A nonmalignant tumor in the uterus, which may prevent the embryo from properly implanting into the endometrium or might cause pain, bleeding, miscarriage, and symptomatic enlargement of the uterus.

fibrous bands Scar tissue that may distort the interior of the uterus and prevent the embryo from implanting properly.

fimbriae Fingerlike protrusions from the ends of the fallopian tubes that retrieve the egg or eggs at the time of ovulation.

follicles Blisterlike structures within the ovary that contain eggs and that produce female sex hormones.

follicle-stimulating hormone (FSH) A gonadotropin that is released by the pituitary gland to stimulate the ovaries or testicles. FSH, when marketed in the United States, is also known as Metrodin.

follicular phase insufficiency or defect An abnormal pattern of estrogen production during the first half of the menstrual cycle, which could result in infertility or recurrent miscarriages (should pregnancy occur).

follicular phase of the menstrual cycle *See* proliferative phase of the menstrual cycle.

fornix (pl. *fornices*) Deep recesses in the upper vagina created by the protrusion of the cervix into the roof of the vagina.

FSH   *See* follicle-stimulating hormone.

gamete   The female egg and the male sperm.

gamete intrafallopian transfer (GIFT)   A therapeutic gamete-related technique that involves the injection of one or more eggs mixed with washed, capacitated, and incubated sperm directly into the fallopian tube(s) in the hope that fertilization will occur in vivo and that a healthy pregnancy will follow.

gamete micromanipulation   A special procedure performed on eggs to promote in vitro fertilization in cases where there is severe sperm dysfunction.

gastrulation   The stage of embryonic development in which blastomeres are dedicated to the development of specific organs and structures.

gestation   The period from conception to delivery.

gestational surrogacy   The performance of in vitro fertilization using the prospective parents' gametes and the subsequent transfer of the embryos into the uterus of a third party who thereon would carry the baby to term.

GIFT   *See* gamete intrafallopian transfer.

GnRH   *See* gonadotropin-releasing hormone.

GnRHa   *See* gonadotropin-releasing hormone agonists.

gonadotropin-releasing hormone agonists (Gn RHa)   GnRH-like hormones that block the body's release of both FSH and LH. Through blocking LH production, GnRH agonists are capable of improving a woman's response to fertility drugs and may be used in combination with fertility hormones to promote an enhanced response in women who demonstrate resistance to controlled ovarian hyperstimulation. In the United States, GnRH agonists are also known as Lupron, Synarel, and Nafarelin.

gonadotropin-releasing hormone (GnRH)   A "messenger hormone" released by the hypothalamus to influence the production of gonadotropins by the pituitary gland.

gonadotropins   The gonad-stimulating hormones LH and FSH, which are released by the pituitary gland to stimulate the testicles in the man and the ovaries in the woman.

gonads   The ovaries and testicles.

gonococcus   A bacterium producing gonorrhea, a common venereal disease occurring in both men and women that may cause sterility.

gonorrhea   A common venereal disease that may cause sterility in both men and women.

growth medium A physiological solution that promotes cleavage and development of the embryo.

hatching Opening of the zona (outer shell of the egg) due to expansion of the volume of the embryo through repeated cleavage. It occurs a few days after the embryo arrives or is deposited in the uterus and immediately precedes implantation (*see also* ASSISTED HATCHING).

hCG *See* human chorionic gonadotropin.

hemi-zona test Used to determine whether sperm are able to attach to or penetrate the surface of human eggs.

heparin A drug that may be added to the solution used to flush eggs out of ovarian follicles during egg retrieval; its purpose is to prevent blood clotting within the fluid that harbors the egg.

HLA antigens The imprints of the man's immunologic make up.

hMG *See* human menopausal gonadotropin.

hormonal insufficiency A condition resulting in infertility and/or miscarriage; in the IVF setting, hormonal insufficiency may be produced by an abnormal response to fertility drugs and may lead to the failure of an embryo to implant because the amount of hormones produced and the timing of their production and release were not perfectly synchronized.

hormone (sex hormone) Chemicals produced by the testicles, ovaries, and adrenal glands that play a major role in reproduction and sexual identity.

HSG *See* hysterosalpingogram.

Hühner Test *See* postcoital test.

human chorionic gonadotropin (hCG) A hormone, produced by the implanting embryo (and subsequently also by the placenta), whose presence in the woman's blood indicates a possible pregnancy; hCG may also be administered to women undergoing stimulation with hMG alone or in combination with other fertility drugs in order to trigger ovulation. Injections of hCG may also be administered to encourage the production of progesterone by the corpus luteum in the hope of promoting implantation following embryo transfer and thereby reducing the incidence of spontaneous miscarriage in a pregnancy resulting from IVF. The hormone hCG is derived from the urine of pregnant women.

human menopausal gonadotropin (hMG) A natural hormone that is administered either alone or in combination with other fertility drugs to induce ovulation of more than one egg. The hormone hMG is derived from the urine of menopausal women. When marketed in the United States, hMG is also known as Pergonal.

hypothalamus A small area in the midportion of the brain that, together with the pituitary gland, regulates the formation and release of many hormones in the body, including estrogen and progesterone by the ovaries and testosterone by the testes.

hysterosalpingogram (HSG) A procedure used to assess the interior of the fallopian tubes and uterus; it involves injecting a dye into the uterus via the vagina and cervix, and tracking the dye's pathway by a series of X rays.

hysteroscope A lighted, telescope-like instrument that is passed through the cervix into the uterus, enabling the surgeon to examine the cervical canal and the inside of the uterus for defects or disease.

hysteroscopy Examination of the cervical canal and inside of the uterus for defects, by means of the hysteroscope. Surgery designed to correct such defects can be performed through the hysteroscope during this procedure, thereby often making more invasive abdominal surgery unnecessary.

ICSI *See* intracytoplasmic sperm injection.

immature oocyte (or egg) retrieval The retrieval of numerous healthy but immature eggs from women who had *not* received any fertility drugs in advance of the egg retrieval; these eggs are subjected to a complex process of maturation in the laboratory and are then fertilized using ICSI.

implantation The process that occurs when the embryo burrows into the endometrium and eventually connects to the mother's circulatory system.

inclusive pregnancy rates Pregnancy success reports that combine rates for both clinical and chemical pregnancies and do not distinguish between the two.

infertility The inability to conceive after one full year of normal, regular heterosexual intercourse without the use of contraception.

insemination In the laboratory, the addition of a drop or two of the medium containing capacitated sperm to a petri dish containing the egg in order to achieve fertilization. Also refers to placement of sperm into the woman's reproductive tract.

insemination medium A liquid that bathes and nourishes the eggs and embryos in the petri dish just as the mother's body fluids sustain them in nature.

intracytoplasmic sperm injection (ICSI) A form of micromanipulation whereby a single sperm is captured in a thin glass needle and injected directly into the ooplasm of the egg. Usually used to assist fertilization in couples suffering from severe sperm dysfunction.

intrauterine insemination (IUI) The injection of sperm, processed in the laboratory, into the uterus by means of a catheter directed through the cervix; enables sperm to reach and fertilize the egg more easily or to bypass hostile cervical mucus.

intravaginal insemination (IVI) The injection of semen (usually donor semen) into the vagina in direct proximity to the cervix in the hope that pregnancy will occur.

invasive procedure Any operative procedure, major or minor, that traverses body tissues. In the case of fertility-related treatments, a surgical procedure that requires that one or more punctures or incisions be made in the woman's abdomen.

in vitro fertilization (IVF or IVF/ET) Literally "fertilization in glass," IVF comprises several basic steps: the woman is given fertility drugs that stimulate her ovaries to produce a number of mature eggs; at the proper time, the eggs are retrieved by suction through a needle that has been inserted into her ovaries; the eggs are fertilized in a glass petri dish, or in a test tube, in the laboratory with her partner's or donor sperm; and subsequently the embryos are transferred into the body.

in vivo fertilization Fertilization inside the body.

isohormones Similarly structured components that have different levels of biological activity; the influence of isohormones may be responsible for the variations in potency among different batches of gonadotropins such as hMG and purified FSH.

IUI *See* intrauterine insemination.

IVF third-party parenting A situation in which an individual other than one of the aspiring parents provides gametes (as with sperm or ovum donation) or a uterus, and the woman who will carry the baby to term undergoes embryo transfer.

IVF or IVF/ET *See* in vitro fertilization.

IVF surrogacy Synonym for gestational surrogacy.

IVI *See* intravaginal insemination.

knee-chest position Position the woman may be asked to assume during embryo transfer if the uterus is tipped forward, to contribute to optimal placement of the embryos.

labia majora The hair-covered outer lips of the external portion of the female reproductive tract.

labia minora The small inner lips of the outer female reproductive tract, partially hidden by the labia majora.

laparoscope A long, thin telescope-like instrument containing a high-intensity light source and a system of lenses that enables the surgeon to examine the abdominal/pelvic cavity and to perform other diagnostic or surgical procedures under direct vision without necessitating major surgery.

laparoscopy A surgical procedure using the laparoscope. Laparoscopy may be used for egg retrieval, diagnostic evaluation, reparative surgery, and various other fertility procedures. Because of its dual abilities to enable the physician to assess tubal patency and visualize the abdominal cavity, laparoscopy has largely replaced hysterosalpingography as the most popular method of assessing the anatomical integrity of the reproductive tract (*see also* augmented laparoscopy). Once the favored procedure for egg retrieval, too, laparoscopic egg retrieval has been supplanted by ultrasound-guided egg retrieval.

laparotomy A procedure in which an incision is made in the abdomen to expose the abdominal contents for diagnosis or surgery.

LH *See* luteinizing hormone.

lithotomy Position that a woman is asked to assume in order to undergo a gynecological examination or other procedure such as embryo transfer, vaginal ultrasound examination, etc.

luteal-phase insufficiency or defect The inadequate production of hormones during the second phase of the menstrual cycle, which may result in infertility or miscarriage.

luteal phase of the menstrual cycle *See* secretory phase of the menstrual cycle.

luteinizing hormone (LH) A gonadotropin released by the pituitary gland to stimulate the ovaries and testicles.

macrophages Cells of the immune system that destroy invading organisms or foreign proteins.

male subfertility Less than optimal sperm quality, including configuration, motility, and count (number produced in a semen specimen), that reduces the chance of conception without completely preventing its spontaneous occurrence.

meiosis The process of reducing and dividing the chromosomes in both the sperm and egg, which occurs immediately prior to and during fertilization.

menopause The period of a woman's life that begins with the total cessation of menstruation, usually between the ages of 40 and 55.

menstrual cycle The time that elapses between menstrual periods. The average cycle is 28 days, with ovulation usually occurring at the midpoint (around the 14th day).

menstruation The monthly flow of blood when pregnancy does not occur; the flow comprises about two-thirds of the endometrium and blood, often including the unfertilized egg or unimplanted embryo.

micromanipulation A term used to describe a variety of mechanical procedures used to promote the entry of sperm into the egg. Also called assisted fertilization.

microorganelles Tiny intracellular factories that provide energy and perform metabolic functions in the egg, where the microorganelles are located largely in the ooplasm.

miscarriage Spontaneous expulsion of the products of conception from the uterus in the first half of pregnancy.

mitosis The identical replication of cells by cleavage; mitosis is the process responsible for the growth and development of all tissues.

mock embryo transfer A trial procedure wherein a thin catheter is introduced via the cervix into the uterine cavity. It is intended to simulate embryo transfer and evaluate the potential for embryo transfer. The percentage of sperm that have a normal vs. abnormal shape, structure, or configuration.

morulla An early phase during which the developing embryo, which contains a large number of blastomeres, resembles a mulberry.

motility (sperm motility) The ability of sperm to move and progress forward through the reproductive tract and fertilize the egg; sperm motility can be assessed microscopically.

multiple pregnancy The presence of more than one gestation within the woman's reproductive tract at the same time.

myceles Microfibers within the cervical mucus that sperm must swim through to reach the uterus; the woman's hormonal environment determines whether the arrangement of the myceles will facilitate or inhibit passage of the sperm. Around the time of ovulation the myceles are arranged in a parallel fashion so that sperm can swim between them in order to reach the uterus; it is believed that capacitation is promoted during that process.

natural cycle IVF A situation in which one or two eggs are harvested from a woman's ovaries during the natural menstrual cycle and are then subjected to IVF and embryo transfer. Success rates are much lower than with conventional IVF.

nonpigmented endometriosis A condition in which endometriotic deposits in the pelvis cannot be seen at the time of laparoscopy or laparotomy because blood pigment has not been deposited in these lesions. The condition is believed to often precede the development of visible lesions.

nucleus Structure in the cell that bears the chromosomes.

oocyte *See* egg.

ooplasm Nurturing material around the nucleus of the egg that contains micoorganelles and nurtures the zygote and embryo after fertilization.

operative laparoscope A laparoscope that has been modified to allow passage of a double-bore needle or surgical instruments through a groove or sleeve adjacent to the instrument (*see also* laparoscope).

organic pelvic disease The presence of structural damage in the pelvis due to trauma, inflammation, tumors, congenital defects, or degenerative disease.

ovarian stroma The ovarian tissue surrounding the follicles that produces hormones.

ovaries Two white, almond-sized structures, the female counterpart of the testicles, that are attached to each side of the pelvis adjacent to the ends of the fallopian tubes; the ovaries both release eggs and discharge sex hormones into the bloodstream.

ovulation The process that occurs when an ovary releases one or more eggs.

ovum *See* egg.

partial zona dissection A mechanical form of assisted hatching in which part of the zona (outer shell of the egg) is dissected away in order to promote hatching.

patency Openness, freedom from blockage (particularly referring to the fallopian tubes).

PCT test *See* postcoital test.

peeling Removal of the corona radiata from the embryo by flushing the embryo through a syringe or pipette, or by microdissection using fine instruments. An embryo often must first be peeled before it is possible to determine whether fertilization and cleavage have occurred.

penis The male external sex organ.

Percoll® A chemical substance through which sperm are passed to enhance fertilization potential.

perineum The outer portion of the fibromuscular wall and skin that separate the anus and rectum from the vagina and vulva.

peristaltic movements of fallopian tubes Mechanism by which the fallopian tubes contract in a purposeful and rhythmical way to transport sperm, eggs, and embryos in a timely manner so as to promote fertilization and, ultimately, implantation.

peritoneal cavity The abdominal cavity that contains pelvic organs, bowel, stomach, liver, kidneys, adrenal glands, spleen, and so on, and is lined by a membrane called the peritoneum.

perivitelline membrane Membrane that separates the ooplasm and nuclear material from the zona pellucida in the human egg.

phrenic nerve Nerve that may be irritated by trapped gas or blood during laparoscopy or following internal bleeding, resulting in subsequent pain in the shoulder, arm, and neck (most commonly on the right side).

pituitary gland A small, grapelike structure hanging from the base of the brain that, together with the hypothalamus, produces and regulates the release of many hormones in the body.

placenta The uterine factory that nourishes the fetus throughout pregnancy and is connected to the baby's navel via the umbilical cord.

placentation Formation and attachment of the placenta to the uterine wall.

plasma membrane Double-layered membrane that envelops the entire sperm.

polycystic ovarian disease Condition in which the ovaries develop multiple small cysts; it is often associated with abnormal or absent ovulation and, accordingly, with infertility.

polyps (uterine) Outgrowths that protrude into the uterus and may cause pain and bleeding or prevent an embryo from implanting.

polyspermia The entry of more than one sperm into an egg during fertilization; this causes the zygote to die or the embryo to divide haphazardly and then die.

postcoital (PCT) test Assessment of the cervical mucus after intercourse to evaluate the quality of the mucus and mucus-sperm interaction; also known as the Hühner Test.

pregnancy-specific glycoprotein (SP-1) A hormone that appears in the blood only a few days later than hCG during pregnancy. May be measured instead of hCG to diagnose pregnancy.

pre-implantation genetics A new diagnostic technology involving chromosome and genetic assessment of the embryo in order to determine its health and potential to develop into a healthy offspring. A procedure currently used to determine the sex of the embryo and to diagnose a variety of genetic disorders.

PROD *See* protease digestion.

progesterone A primary female sex hormone produced by the corpus luteum that induces secretory changes in the glands of the endometrium. Progesterone may also be given by injection or in the form of vaginal suppositories to enhance implantation and reduce the risk of miscarriage.

prolactin A hormone produced by the brain that may influence the activity of FSH on the ovaries.

proliferative phase of the menstrual cycle Usually, the first half of the menstrual cycle, when the endometrium grows under the influence of estrogen and the follicles develop; also known as the follicular phase.

prolonged coasting The discontinuation of hMG/FSH medication and deferring hCG administration for a number of days, while continuing GnRH agonist therapy in cases where severe ovarian hyperstimulation occurs following COH in preparation for IVF.

prostaglandins Natural hormones contained in a multitude of cells in the body as well as in the seminal fluid. The placement of semen, which contains seminal fluid with prostaglandins, directly in the uterus in quantities greater than 0.2 ml can cause life-threatening shock.

prostate gland Gland in the male reproductive tract that secretes a milky substance that nurtures and promotes survival of sperm. The combination of sperm and milky fluid that is ejaculated during erotic experiences is known as *semen.*

protease digestion (PROD) In our setting we've developed a new method of assisted hatching using an enzyme (protease XIV), which gently softens and thins the exterior of the zona in its full circumference in a manner analogous to natural hatching.

purified FSH A fertility hormone derived by processing and purifying hMG to eliminate the LH component.

quantitative Beta hCG blood pregnancy test Test that detects and measures the amount of hCG (produced by an implanting embryo) in the woman's blood. Measured nine to 11 days after embryo transfer, it can diagnose a possible pregnancy before the woman has missed a menstrual period.

rectum Lower portion of the large intestine that connects to the anal canal.

Resolve, Inc. One of the largest and most reputable fertility support groups in the United States; its national office is in Somerville, MA.

retrograde ejaculation A condition, sometimes caused by a spinal cord injury or following removal of a diseased prostate gland, in which the man ejaculates backward into the bladder rather than outward through the penis. It may cause infertility but can be treated by inseminating the woman with sperm separated from urine the man would pass immediately following orgasm.

salpingoscopy A procedure involving the introduction of a thin fiber-optic instrument into the fallopian tube or tubes to promote visualization of the tubal lining. It is usually performed during laparoscopy but may also be performed through a hysteroscope.

salpingostomy A form of tubal surgery in which the end of the fallopian tube(s) is opened at the time of laparoscopy or laparotomy, using small surgical stitches or a laser.

SART *See* The Society of Assisted Reproductive Technology.

scrotum Pouch in which the male's testicles are suspended outside the body.

secretory phase of the menstrual cycle The second half of the menstrual cycle, which begins after ovulation under the influence of estrogen and progesterone produced by the corpus luteum; the term *secretory* is derived from the secretion by the endometrium of nutrients that will sustain an embryo; also known as the *luteal phase.*

selective reduction of pregnancy Prior to completion of the third month of pregnancy, reduction of the number of fetuses in a large multiple pregnancy by injecting a chemical substance under ultrasound guidance; the fetus or fetuses succumb almost immediately and are absorbed by the body. It may be considered a life-saving measure for the remaining fetuses in high-multiple pregnancies such as quadruplets, quintuplets, or greater, and may reduce the risk of high-multiple pregnancies.

semen The combination of sperm, seminal fluid, and other male reproductive secretions.

seminal fluid Milky fluid produced by the seminal vesicles that is ejaculated during erotic experiences (*see also* semen).

seminal vesicles Glands in the male reproductive tract that secrete a milky substance that nurtures and promotes survival of sperm.

Society for Assisted Reproductive Technology (SART) Affiliated with the American Society for Reproductive Medicine, SART provides information about IVF programs and compiles a registry of audited IVF results from participating programs.

sperm The male gamete; spermatozoa.

sperm antibody test Test that determines whether either partner's blood or the woman's cervical mucus contains antibodies to sperm.

sperm count A basic fertility-assessment test of sperm function, primarily involving counting the number of sperm, assessing their motility and progression, and evaluating their overall structure and form.

spontaneous menstrual abortion An early miscarriage occurring at the time of menstruation without the woman's menstrual period being delayed.

SP-1 *See* pregnancy-specific glycoprotein.

sterility *See* infertility.

stimulation Induction of the development of a number of follicles in response to the administration of fertility drugs (*see also* controlled ovarian hyperstimulation and superovulation).

subzonal insertion (SUZI) The direct injection of one or more sperm into the perivitelline space to promote fertilization. It is a form of micromanipulation.

superovulation The ovulation of more than one egg induced through the administration of fertility drugs (*see also* controlled ovarian hyperstimulation and stimulation).

surrogation Situation in which an infertile woman uses someone else's uterus to carry a child to term for her. Surrogation can be divided into (1) cases in which the surrogate mother contributes biologically to the offspring by providing her own eggs (classic surrogacy), and (2) cases in which the surrogate does not contribute biologically and therefore must undergo IVF (gestational surrogacy).

SUZI *See* subzonal insertion.

syphilis A life-endangering venereal disease that in its late stages attacks most systems in the body, including the cardiovascular and central nervous systems.

testes *See* testicles.

testicles The male counterparts of the female ovaries; located in the scrotum, the testicles produce sperm and male hormones such as testosterone.

testosterone The predominant male sex hormone, which influences the production and maturation of sperm.

test yolk buffer A sperm-enhancement solution derived from the yolk of a chicken egg.

therapeutic gamete-related technologies Procedures involving the use of gametes to enhance the chance of conception through subsequent insemination, or transfer of eggs and/or sperm into the woman's uterus, fallopian tubes, or peritoneal cavity.

therapeutic hysteroscopy A procedure, usually performed under general anesthesia, in which a clear liquid is injected into the uterine cavity via the hysteroscope. This permits exposure of surface lesions inside the uterus or cervix via X ray. These lesions can be treated surgically through excision, obliteration, transection, etc.

third-party parenting Refers to any situation in which an individual other than the aspiring parents assists by providing gametes (as with sperm and ovum donation) or a uterus in order to help the couple have a baby.

thyroid-stimulating hormone (TSH) A hormone produced by the pituitary gland that stimulates the release of thyroid hormone by the thyroid gland.

TPI *See* transperitoneal insemination.

transabdominal egg retrieval An ultrasound-guided egg retrieval procedure in which the needle is passed through the abdominal wall and a full bladder into the ovarian follicles; has largely been supplanted by transvaginal egg retrieval.

transcervical Refers to the physiological or surgical pathway whereby secretions, organisms, sperm, or surgical instrumentation passes from the vagina into the uterus.

transmyometrial embryo transfer A procedure that transfers embryos to the uterus via a needle and/or catheter introduced through the uterine wall (myometrium) rather than through the cervix. It is used in situations where severe narrowing of the cervix negates the performance of conventional embryo transfer.

transperitoneal insemination (TPI) The injection of washed sperm through a syringe into the woman's pelvic cavity at the time of expected ovulation to promote conception; may be combined with intrauterine insemination.

transurethral egg retrieval An ultrasound-guided egg retrieval procedure in which the needle is passed through the urethra and the bladder wall into the ovaries; has largely been supplanted by transvaginal egg retrieval.

transvaginal egg retrieval An ultrasound-guided egg retrieval procedure in which the needle is passed through the back or side of the woman's vagina into her ovaries. It is the most commonly performed egg retrieval procedure today.

treatment cycle The menstrual cycle during which a particular fertility treatment such as IVF, IUI, AID, GIFT, etc. was performed.

tubal embryo transfer (TET) A procedure, usually performed via laparoscopy, in which one or more embryos are inserted in the fallopian tube(s) via a thin catheter a few days following egg retrieval.

tubal pregnancy *See* ectopic pregnancy.

tubal reanastomosis A surgical procedure in which the fallopian tubes are reconnected, reestablishing patency. Usually performed after a previous tubal ligation (sterilization).

ultrasound A painless diagnostic procedure that transforms high-frequency sound waves as they travel through body tissue and fluid into images on a TV-like screen; it enables the physician to clearly identify structures within the body and to guide instruments during certain procedures. Ultrasound is also used to diagnose a clinical pregnancy.

unexplained infertility Infertility whose cause cannot be readily determined by conventional diagnostic procedures; this occurs in about 10% of all infertile couples.

ureaplasma A microorganism that occurs in the reproductive tracts of males and females, and might interfere with sperm transport and/or embryo implantation. It might also be responsible for early miscarriages.

urethra The canal-like structure through which urine passes from the bladder and through which semen passes during ejaculation.

urine ovulation test A simple test that can pinpoint the time of presumed ovulation; frequent charting of the test results detects the surge of LH that triggers ovulation.

uterus A muscular organ that enlarges during pregnancy from its normal pearlike size to accommodate a full-term pregnancy.

vagina The narrow passage that leads from the vulva to the cervix. The vagina's elastic tissue, muscle, and skin have enormous ability to stretch so as to accommodate the penis during the sex act and the passage of a baby during childbirth.

varicocele A collection of dilated veins around the testicles that hinders sperm function, possibly through increasing the temperature in the scrotum.

vas deferens Tube that connects the epididymis with the urethra in the male reproductive tract.

vasectomy Surgery to block the male's sperm ducts for the purpose of birth control.

vestibule The cleft between the labia minora; the entrance to the vagina.

vulva The external portion of the female reproductive tract.

washing (sperm washing) The processing of a semen specimen in a centrifuge in order to separate the sperm from the semen specimen.

yellow body *See* corpus luteum.

ZIFT *See* zygote intrafallopian transfer.

zona–cumulus complex The mass of cells (zona pellucida and cumulus mass) through which the sperm must pass to reach the egg.

zona drilling A form of micromanipulation whereby a small hole is made in the zona pellucida to promote free entry of a sperm into the egg.

zona-free hamster-egg penetration test A technique that helps determine whether sperm are likely to fertilize healthy eggs.

zona pellucida The shell-like covering of the human egg.

zygote The term for a fertilized egg until it begins to cleave, at which time it is known as an *embryo*.

zygote intrafallopian transfer (ZIFT) The placement of one or more zygotes into the outer third of the fallopian tube(s) during laparoscopy or minilaparotomy in the hope that the resulting embryo(s) will travel to the uterus and implant successfully.

Index

Italic page numbers indicate illustrations or captions.
Page numbers followed by a "g" indicate glossary terms.

A

abortion 47, 173
accreditation 187–189
acid tyrodes digestion 194g
acquired immune deficiency syndrome
(AIDS) *see* AIDS
acrosome *13*, 14, 194g
acrosome reaction 17, 194g
adenomyosis 98, 194g
adrenal glands 194g
adrenalin 78
AF *see* assisted fertilization
age
  egg quality related to 93–94
  implantation success related to 87–88
  risks associated with 100–101
AID *see* artificial insemination by donor
AIDS (acquired immune deficiency
syndrome) 42, 169, 194g
ALA *see* antilymphocyte antibodies
alcohol 4, 28
Allenson, Sandra 147
allergic response 32
allo-immune response 90
alloimmunity 194g
alphafeto protein 37, 194g
American College of Obstetricians and
Gynecologists 180, 187
American Fertility Society *see* American
Society for Reproductive Medicine
American Medical Association (AMA) 180
American Society for Reproductive
Medicine (ASRM) (formerly American
Fertility Society) xv, 180, 187, 194g
amniocentesis 37
anatomy, reproductive 6–28
Andereck, William 167
androgens 58

anesthetics 65–66 *see also* paracervical
block
antibodies 32, 36, 89–93 *see also* sperm
antibodies
antileucocyte antibodies *see*
antilymphocyte antibodies (ALA)
antilymphocyte antibodies (ALA) 90, 92,
195g
antiphospholipid antibodies (APA) 90, 91,
195g
anus *7, 7, 8,* 195g
APA *see* antiphospholipid antibodies
ART *see* assisted reproductive technology
artificial insemination 146–149
artificial insemination by donor (AID) 149,
195g
Asch, Ricardo 150
aspirin 91, 92
ASRM *see* American Society for
Reproductive Medicine
assisted fertilization (AF) 88, 176–178,
195g *see also* micromanipulation
assisted hatching 178–179, 195g
assisted reproductive technology (ART)
195g
augmented laparoscopy 195g
Austria 82
autoantibodies 196g
auto-immune hypothyroidism *see*
Hashimoto's disease
auto-immunity 90–91

B

baking soda (sodium bicarbonate) 71
basal body temperature (BBT) chart
107–108, *109,* 196g
Billings Method of contraception 20, 196g
biochemical pregnancy 27